Milankovitch and Climate

Understanding the Response to Astronomical Forcing

NATO ASI Series
Advanced Science Institutes Series

A series presenting the results of activities sponsored by the NATO Science Committee, which aims at the dissemination of advanced scientific and technological knowledge, with a view to strengthening links between scientific communities.

The series is published by an international board of publishers in conjunction with the NATO Scientific Affairs Division

A	Life Sciences	Plenum Publishing Corporation
B	Physics	London and New York
C	Mathematical and Physical Sciences	D. Reidel Publishing Company Dordrecht, Boston and Lancaster
D	Behavioural and Social Sciences	Martinus Nijhoff Publishers
E	Engineering and Materials Sciences	The Hague, Boston and Lancaster
F	Computer and Systems Sciences	Springer-Verlag
G	Ecological Sciences	Berlin, Heidelberg, New York and Tokyo

Series C: Mathematical and Physical Sciences Vol. 126 Part 2

Milankovitch and Climate

Understanding the Response to Astronomical Forcing

edited by

A. Berger
Institut d'Astronomie et de Géophysique Georges Lemaitre,
Université Catholique de Louvain-la-Neuve, Belgium

J. Imbrie
Department of Geological Sciences,
Brown University, Providence, Rhode Island, U.S.A.

J. Hays
Lamont-Doherty Geological Observatory,
Columbia University, Palisades, New York, U.S.A.

G. Kukla
Lamont-Doherty Geological Observatory,
Columbia University, Palisades, New York, U.S.A.

B. Saltzman
Department of Geology and Geophysics,
Yale University, New Haven, Connecticut, U.S.A.

D. Reidel Publishing Company

Dordrecht / Boston / Lancaster

Published in cooperation with NATO Scientific Affairs Division

Proceedings of the NATO Advanced Research Workshop on
Milankovitch and Climate
Palisades, New York, U.S.A.
November 30 – December 4, 1982

Library of Congress Cataloging in Publication Data

Milankovitch and climate.

(NATO ASI Series. Series C, Mathematical and physical sciences; 126)
"Proceedings of the NATO Advanced Research Workshop on Milankovitch and Climate, Lamont-Doherty Geological Observatory, Palisades, N.Y., November 30-December 4, 1982, CIP t.p. verso.
Includes index.
1. Paleoclimatology—Congresses. 2. Astrophysics—Congresses. 3. Milankovitch, Milutin. I. Berger, A. (Andre), 1942-ñ . II. NATO Advanced Research Workshop on Milankovitch and Climate (1983: Lamont-Doherty Geological Observatory) III. Series.
QC884.M55   1984   551.6   84-6805
ISBN 90-277-1777-X (Part 1)
ISBN 90-277-1778-8 (Part 2)
ISBN 90-277-1791-5 (Set)

Published by D. Reidel Publishing Company
P.O. Box 17, 3300 AA Dordrecht, Holland

Sold and distributed in the U.S.A. and Canada
by Kluwer Academic Publishers,
190 Old Derby Street, Hingham, MA 02043, U.S.A.

In all other countries, sold and distributed
by Kluwer Academic Publishers Group,
P.O. Box 322, 3300 AH Dordrecht, Holland

D. Reidel Publishing Company is a member of the Kluwer Academic Publishers Group

All Rights Reserved
© 1984 by D. Reidel Publishing Company, Dordrecht, Holland.
No part of the material protected by this copyright notice may be reproduced or utilized in any form or by any means, electronic or mechanical, including photocopying, recording or by any information storage and retrieval system, without written permission from the copyright owner.

Printed in The Netherlands.

TABLE OF CONTENTS

CONTENTS OF PART 1   vii

PART III - MODELLING LONG-TERM CLIMATIC VARIATIONS IN RESPONSE TO ASTRONOMICAL FORCING

Section 1 - Energy Balance Climate Models............... 511

NORTH G.R., J.G. MENGEL and D.A. SHORT - A two-dimensional climate model useful in ice age applications.. 513
ELIASEN E. - A simple seasonal climate model and its response to variations in the orbital parameters..... 519
ADEM J., A. BERGER, Ph. GASPAR, P. PESTIAUX and J.P. van YPERSELE - Preliminary results on the simulation of climate during the last deglaciation with a thermodynamic model.. 527

Section 2 - Models with an ice sheet................... 539

POLLARD D. - Some ice-age aspects of a calving ice-sheet model.. 541
PELTIER W.R. and W. HYDE - A model of the ice age cycle. 565
LEDLEY T.S. - Sensitivities of cryospheric models to insolation and temperature variations using a surface energy balance...................................... 581
WATTS R.G. and Md.E. HAYDER - A possible explanation of differences between Pre- and Post-Jaramillo ice sheet growth... 599
BIRCHFIELD G.E. and J. WEERTMAN - On long period internal oscillations in a simple ice sheet model......... 605
OERLEMANS J. - On the origin of the ice ages............ 607

Section 3 - Oscillator Models of Climate................. 613

SALTZMAN B., A. SUTERA and A.R. HANSEN - Earth-orbital eccentricity variations and climatic change.......... 615
NICOLIS C. - Self-oscillations, external forcings, and climate predictability................................ 637
HARVEY L.D.D. and S.H. SCHNEIDER - Sensitivity of internally-generated climate oscillations to ocean model formulation... 653

Section 4 - Conceptual Models of Climatic Response...... 669

RUDDIMAN W.F. and A. McINTYRE - An evaluation of ocean-climate theories on the North Atlantic............... 671
BROECKER W.S. - Terminations............................ 687
COVEY C. and S.H. SCHNEIDER - Models for reconstructing temperature and ice volume from oxygen isotope data.. 699
YOUNG M.A. and R.S. BRADLEY - Insolation gradients and the paleoclimatic record............................ 707
JOHNSON R.G. - Summer temperature variations in the Arctic during interglacials............................ 715
FOLLAND C. and F. KATES - Changes in decadally averaged sea surface temperature over the world 1861-1980..... 721
FAIRBRIDGE R.W. - Sea-level fluctuations as evidence of the Milankovitch cycles and of the planetary-solar modulation of climate................................ 729

Section 5 - General Circulation Models................. 731

ROYER J.F., M. DEQUE and P. PESTIAUX - A sensitivity experiment to astronomical forcing with a spectral GCM : simulation of the annual cycle at 125 000 BP and 115 000 BP...................................... 733
SELLERS W.D. - The response of a climate model to orbital variations...................................... 765
MANABE S. and A.J. BROCCOLI - Influence of the CLIMAP ice sheet on the climate of a general circulation model : implications for the Milankovitch theory..... 789
KUTZBACH J.E. and P.J. GUETTER - The sensitivity of monsoon climates to orbital parameter changes for 9 000 years BP : experiments with the NCAR general circulation model.. 801

PART IV - CLIMATIC VARIATIONS AT ASTRONOMICAL FREQUENCIES. Summary, Conclusions and Recommendations 821

AUTHORS INDEX... 875
SUBJECTS INDEX.. 887

CONTENTS OF PART 1

INTRODUCTION..
LIST OF PARTICIPANTS......................................
MILANKOVITCH V. - The Memory of my Father ....

PART I - ORBITAL AND INSOLATION VARIATIONS

BERGER A. - Accuracy and frequency stability of the Earth's orbital elements during the Quaternary......
BRETAGNON P. - Accuracy of long term planetary theory...
BUYS M. and M. GHIL - Mathematical methods of celestial mechanics illustrated by simple models of planetary motion..
BERGER A. and P. PESTIAUX - Accuracy and stability of the Quaternary terrestrial insolation................
TAYLOR K.E. - Fourier representations of orbitally induced perturbations in seasonal insolation..........

PART II - GEOLOGICAL EVIDENCE FOR LONG-TERM CLIMATIC VARIATIONS AT ASTRONOMICAL FREQUENCIES

Section 1 - Pre-Pleistocene evidence of orbital forcing

OLSEN P.E. - Periodicity of lake-level cycles in the Late Triassic Lockatong formation of the Newark basin (Newark supergroup, New Jersey and Pennsylvania......
ANDERSON R.Y. - Orbital forcing of evaporite sedimentation...
FISCHER A.G. and W. SCHWARZACHER - Cretaceous bedding rhythms under orbital control ? ....
de BOER P.L. and A.A.H. WONDERS - Astronomically induced rhythmic bedding in Cretaceous pelagic sediments near Moria (Italy)...
ARTHUR M.A., W.E. DEAN, D. BOTTJER and P.A. SCHOLLE - Rhythmic bedding in Mesozoic-Cenozoic pelagic carbonate sequences : the primary and diagenetic origin of Milankovitch-like cycles...........................

FILLON R.H. - Ice-age Arctic ocean ice-sheets : a possible direct link with insolation ....
HERMAN Y. and J.K. OSMOND - Late Neogene Arctic paleoceanography : micropaleontology and chronology.......
PANTIĆ N. and D. STEFANOVIĆ — Complex interaction of cosmic and geological events that affect the variation of Earth climate through the geologic history...
DEAN W.E. and J.V. GARDNER - Cyclic variations in calcium carbonate and organic carbon in Miocene to Holocene sediments, Walvis Ridge, South Atlantic ocean...

Section 2 - Marine Pleistocene Records of Climatic Response..

IMBRIE J., J.D. HAYS, D.G. MARTINSON, A. McINTYRE, A.C. MIX, J.J. MORLEY, N.G. PISIAS, W.L. PRELL and N.J. SHACKLETON - The orbital theory of Pleistocene climate : support from a revised chronology of the marine $\delta^{18}O$ record..
PISIAS N.G. and M. LEINEN - Milankovitch forcing of the oceanic system : evidence from the northwest Pacific.
JANECEK T.R. and D.K. REA - Pleistocene fluctuations in northern hemisphere tradewinds and westerlies........
PRELL W.L. - Monsoonal climate of the Arabian sea during the Late Quaternary : a response to changing solar radiation...
ROSSIGNOL-STRICK M. - Immediate climate response to orbital insolation : Mediterranean sapropels and the African monsoon...

Section 3 - Non-Marine Records of Pleistocene Climate...

HOOGHIEMSTRA H. - A Palynological registration of climatic change of the last 3.5 million years..........
AHARON P. - Implications of the coral-reef record from New Guinea concerning the astronomical theory of ice ages...
MOLFINO B., L.H. HEUSSER and G.M. WOILLARD - Frequency components of a Grande Pile pollen record : evidence of precessional orbital forcing.....................
KANARI S., N. FUJI, and S. HORIE - The Paleoclimatological constituents of paleotemperature in lake Biwa....

TABLE OF CONTENTS

Section 4 - Estimation of Geologic Spectra..............

PESTIAUX P. and A. BERGER - An optimal approach to the spectral characteristics of deep-sea climatic records..

HERTERICH K. and M. SARNTHEIN - Brunhes time scale : tuning by rates of calcium-carbonate dissolution and cross spectral analyses with solar insolation........

MORLEY J.J. and N.J. SHACKLETON - The effect of accumulation rate on the spectrum of geologic time series : evidence from two south atlantic sediment cores......

DALFES H.N., S.H. SCHNEIDER and S.L. THOMPSON - Effects of bioturbation on climatic spectra inferred from deep-sea cores..................................

PESTIAUX P. and A. BERGER - Impacts of deep-sea processes on paleoclimatic spectra........................

PART III

MODELLING LONG-TERM CLIMATIC VARIATIONS IN RESPONSE TO ASTRONOMICAL FORCING

SECTION 1 - ENERGY BALANCE CLIMATE MODELS

A TWO-DIMENSIONAL CLIMATE MODEL USEFUL IN ICE AGE APPLICATIONS

G.R. North[1], J.G. Mengel[2], D.A. Short[1]

[1]Goddard Laboratory for Atmospheric Sciences, NASA/Goddard Space Flight Center, Greenbelt, Maryland 20771, USA

[2]Applied Research Corporation, Landover, Maryland 20785, USA

In recent years there have been many attempts to use simple energy balance models to simulate the climatic changes that occurred during the Pleistocene by altering the earth's energy budget through small orbital element changes. Attempts with mean annual models which resolved latitudinal variation in temperature failed to produce large ice sheets when the obliquity was changed a few degrees (1,2) although these early models were generally more sensitive than today's parameterizations would suggest. Attention then turned to seasonal energy balance models, since the primary insolation anomaly is seasonal. Sellers (3), Thompson and Schneider (4) and North and Coakley (5) developed simple models which were either effectively zonally averaged or included a separate land and ocean surface temperature at each latitude. Suarez and Held (6) added some vertical resolution in their model. All included the ice-albedo feedback mechanism and horizontal transport which was essentially diffusive. The models shared one interesting property : they remained rather insensitive to the sizes of orbital changes which were thought to have occurred in the Pleistocene. There were minor differences in the results and some pointed toward possible ways out of the puzzle but curiously the mechanisms of ice--albedo feedback in the insolation and water vapor feedback in the infrared terms seemed too weak to account for the ice ages (for a review of these models see (7) and more recently (8)).

All of the cited results are equilibrium studies; i.e., the orbital element is changed and an equilibrium seasonal cycle is computed assuming the ice sheets are in balance. It has now become apparent from the data that large phase lags exist among ice volume, surface temperature and insolation changes (e.g. (9)). This non-equilibrium nature of the ice volume makes the puzzle cited in the first paragraph even more baffling since the equilibrium ice sheets should be much larger than the observed ones (cf. e.g., (10), (8)). Could it be that the dynamics of the glacial flow actually enter the process to enhance the sensitivity?

The nonlinearity and longer time scales associated with large continental ice sheets and the isostatic rebound of bedrock form a natural starting point for nonequilibrium studies. There might also be some hope that the heretofore ignored 100 kyr cycle might emerge as a combination tone in the nonlinear mixer (11). Although the latter may find its explanation in these mechanisms, it seems that global temperature changes remain very small (<1.0K) in the rather complete model put forth by Pollard (12); however, in a very similar approach Watts and Hayder (13) report a somewhat larger response. On the other hand when Pollard uses a concentration of atmospheric CO_2 about one half the present value, the sensitivity problem vanishes. Since this latter assumption is supported by empirical evidence (14), the long sought after missing feedback may be in hand (see also the paper by Broecker at this conference). This particular "way out" of the missing sensitivity puzzle also solves the problem of why the glaciations in the two hemispheres seem to be synchronous.

The major triumph of the models of Pollard (12) and of Watts and Hayder (13) is that they seem capable of reproducing the rather asymmetric growth and decay characteristics of the 100kyr cycle. We feel that this is a significant advance in our understanding of the great glacial fluctuations.

In both of these approaches a glacier model of the Weertman (15) type (cf. also Oerlemans, 16) is embedded in a seasonal climate model of the North-Coakley type. We wish to report here an extension of the North-Coakley model which should prove useful in future nonequilibrium studies along the lines discussed above. The details of the model and a comparison with present seasonal data is being presented elsewhere (17). We shall present here a brief summary and discuss the implications.

The model is described by an energy equation

$$C(\hat{r})\partial T(\hat{r},t)/\partial t - \nabla \cdot D(x) T(\hat{r},t) + A + BT(\hat{r},t) = QS(x,t)a(x,T(\hat{r},t)) \quad [1]$$

where $\hat{r}$ = point on the earth's surface,

t = time of the year (years).
T = temperature (°C) at $\hat{r}, t$
x = sine of latitude.
D(x) = isotropic thermal diffusion coefficient at x
A + BT = empirical IR rule (Wm^{-2})
Q = solar constant
a = coalbedo containing a smooth latitudinal dependence and a hard ice-albedo edge at points on the earth which never go above freezing.
C($\hat{r}$) = effective local heat capacity.

The physical significance and limitations of these terms has been discussed by many authors (e.g., (7)), and need not concern us here. The model is, of course, highly idealized and results should not be taken literally. The land-sea geography is introduced only in the heat capacity function C(r) which is characteristic of the mixed layer over ocean (75m) and of a column of air over land (60 times smaller); over perennial sea ice an intermediate value was chosen to best fit the present climate.

The model mimics the two dimensional distribution of the amplitude and phase of the first and second seasonal harmonics with rather surprising verisimilitude considering the paucity of adjustable parameters used. A crude ice cap rule was adopted : if the temperature never goes above freezing we consider that a (nearly white) ice sheet is present. This simple accounting device allows us to have ice-sheet changes without consideration of the moisture budget. Naturally, this type of rule cannot give us volume nor can it allow us to follow the temporal development of ice-volume. According to this rule Greenland and Antarctica are covered by ice-sheets in the present climate. The model is fairly insensitive to small solar constant changes with $\Delta T \approx 1.5°C$ for a one percent change in Q.

The ice sheet size is sensitive to changes in the summer insolation near the ice sheet edge in this model. By utilizing the orbital parameters of 115 kyr BP we can make summers in the arctic cooler than at present since smaller obliquity coincides with northern summer at aphelion in conjunction with larger eccentricity.

Figure 1 shows a difference map for model July between the present and 115 kyr BP with the ice-albedo feedback suppressed. Clearly the large temperature change in the Northern Canada region corresponds with the ice line's present location whereas the corresponding large temperature change in Asia is well south of the ice line and will have little effect upon ice-sheet growth. This coincidence due to a contrasting land-sea configuration in the Western Hemisphere is much more favorable to ice sheet advances than in the Eastern Hemisphere which is dominated by a single large land mass. Surely moisture budgets also play a role, but the present study suggests that continentality alone is sufficient to favor western hemisphere glaciation.

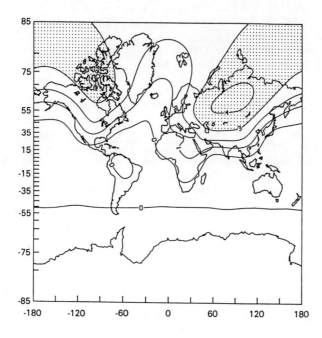

Figure 1 Difference map. Present July minus cool summer orbit July for the linear model. Values greater than 3°C are stippled.

Figure 2 shows a map of the equilibrium ice sheet for orbital conditions 115 kyr BP, supporting the argument just given. As the orbital parameters are slowly changed to favor cool summers, an abrupt transition to the large asymmetrical ice cap occurs. The details of this transition may or may not be physical and we consider it conjectural at this point. The model with large ice cap has a global average temperature about 1°C cooler than at present. This change gives a rough idea of the rather mild ice-albedo effect in the model. Like the earlier studies mentioned above we still need the help of a boost in sensitivity which CO_2 shifts could provide ($\simeq 2°C$) (cf., 18).

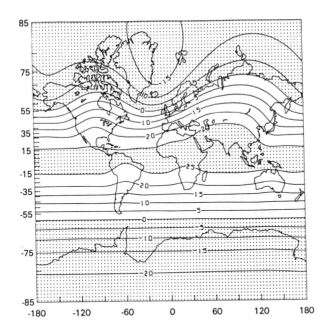

Figure 2 Ice-feedback model July temperature field (°C) for the cool summer orbit. Values less than zero and greater than 25 are stippled. The model infers perennial ice-sheets poleward of the 0°C line in the Northern Hemisphere.

CONCLUSIONS

The present model which is easily and efficiently solved should prove useful for future studies coupling a simple climate model with a deluxe ice sheet model. The climate model already has the property that glaciers have the potential for growing in the right places and with a reasonable sensitivity (especially with the assistance of CO_2 changes). The model in its present form has some rich mathematical structure associated with geography leading to abrupt transitions in ice area as parameters are slowly varied. This feature deserves more attention and we intend to study it in both larger and smaller versions of the model.

REFERENCES

1. Sellers, W.D. : 1970, J. Appl. Met. 9, pp. 960-961.
2. Saltzman, B., and Vernekar, A. : 1971, J. Geophys. Res. 76, pp. 4194-4197.

3. Sellers, W.D. : 1973, J. Appl. Met. 12, pp. 241-254.
4. Thompson, S., and Schneider, S.H. : 1979, J. Geophys. Res. 84, pp. 2401-2414.
5. North, G.R., and Coakley, J.A. : 1979, J. Atmos. Sci. (36), pp. 1189-1204.
6. Suarez, M., and Held, I. : 1979, J. Geophys. Res. 84, pp. 4825-4836.
7. North, G.R., Cahalan, R.F., and Coakley, J.A. : 1981, Rev. Geophys. and Sp. Phys. 19, pp. 91-121.
8. Held, I. : 1982, Icarus 50, pp. 449-461.
9. Ruddiman, W., and McIntyre, A. : 1981, Science 212, pp. 317-627.
10. Imbrie, J., and Imbrie, J.Z. : 1980, Science 207, pp. 943-953.
11. Birchfield, G.E. : 1977, J.Geophys. Res. 82, pp.4909-4913.
12. Pollard, D. : 1983, J. Geophys. Res., in press.
13. Watts, R., and Hayder, E. : 1983, Tellus (submitted).
14. Delmas, R.J., Ascencio J.M., and Legrand, M. : 1980, Nature 284, pp. 155-157.
15. Weertman, J. : 1976, Nature 261, pp. 17-20.
16. Oerlemans, J. : 1980, Nature 287, pp. 430-432.
17. North, G.R., Mengel, J.G., and Short, D.A. : 1983, J. Geophys. Res. 88, pp. 6576-6586.
18. North, G.R., Mengel, J.G., and Short, D.A. : 1983, to be included in proceedings of the IVth Ewing Symposium, published by American Geophysical Union.

A SIMPLE SEASONAL CLIMATE MODEL AND ITS RESPONSE TO VARIATIONS IN THE ORBITAL PARAMETERS

E. Eliasen

Institute of Theoretical Meteorology, University of Copenhagen, Copenhagen, Denmark.

INTRODUCTION

The energy balance models of the type first formulated by Budyko (1) and Sellers (2) are indeed very simple. They contain only one variable quantity, namely the zonally averaged surface temperature, and all physical processes represented in the model have to be parameterized in terms of this temperature field. In order to improve the original simple models it seems natural to represent the temperature in the atmosphere by more than one horizontal field. This should make possible a more accurate treatment of the physical processes, like radiation and convection. Furthermore it should also make it possible to develop a more advanced parameterization of the important meridional and vertical heat transport connected with the general circulation of the atmosphere. Such more complex energy balance models have been formulated by Suarez and Held (3) and by Eliasen and Laursen (4). A variant of the model formulated by Eliasen and Laursen is described in the following and it is shown, that the model has a significant sensitivity to the changes of the distribution of solar radiation caused by variations in the orbital parameters.

THE MODEL

The temperature of the troposphere is represented by the zonally averaged temperature fields T_1 and T_3 at the pressure levels p_1=400 mb and p_3=800 mb, respectively. Using the parameterization of the large-scale heat transport in terms of an equivalent meridional circulation as proposed by Eliasen (5), the two temperature fields as functions of time t and latitude ϕ are determined from the thermodynamic equation by the two equations

$$\frac{\partial T_1}{\partial t} + E\{ r_1 \sigma(\theta_2 - [\theta_2]) - q_1 \frac{\partial \beta}{\partial \phi} \frac{\partial T_1}{\partial \phi}\} = \frac{Q_1}{c_p}$$

$$\frac{\partial T_3}{\partial t} + E\{ r_3 (\theta_2 - [\theta_2]) + q_3 \frac{\partial \beta}{\partial \phi} \frac{\partial T_3}{\partial \phi}\} = \frac{Q_3}{c_p}$$

[1]

where $\theta_2 = 1/2(\theta_1 + \theta_3)$ and $\sigma = 1/2(\theta_1 - \theta_3)$. θ denotes the potential temperature and

$$T = r\theta, \quad r = (\frac{p}{p_*})^\kappa$$

where $p_* = 1000$ mb and $\kappa = R/c_p$ with R being the gas constant and c_p the specific heat capacity of air at constant pressure. The symbol [ ] indicates the horizontal mean value over the whole sphere. The field β may be considered as a smoothed temperature field, defined by the relation

$$\frac{\partial (\cos \phi \frac{\partial \beta}{\partial \phi})}{\cos \phi \, \partial \phi} = [\theta_2] - \theta_2, \quad [\beta] = 0.$$

[2]

The parameterization of the large-scale horizontal as well as vertical heat transport is expressed by using only one coefficient, E. A possible dependence of E upon the global parameters of the system was discussed in Eliasen (5), where the normal value $E=3.1 \cdot 10^{-8} K^{-1} s^{-1}$ was given, a value also used in the present application of the model. Furthermore the values of q_1 and q_3 were given as

$$q_1 = 1 + \frac{r_3 - r_1}{2r_1} = 1.1097, \quad q_3 = 1 - \frac{r_3 - r_1}{2r_3} = 0.9100,$$

where the deviations from the value 1 are due to a crude incorporation of the frictional heating in the dynamical heating. Thus, the quantities Q_1 and Q_3 on the right hand side of eqs.[1] represent the heating rates per unit mass for the part of the heating due to the radiation processes and to the small-scale vertical heat transport. In terms of the absorbed solar radiation A, the upward net fluxes of long-wave radiation F and small-scale heat transport H, the heating rates may be written

$$Q_1 = \frac{g}{\Delta P}(A_1 + F_M - F_T + H_M),$$
$$Q_3 = \frac{g}{\Delta P}(A_3 + F_S - F_M + H_S - H_M),$$

[3]

where g is the acceleration of gravity, $\Delta p = 400$ mb and where the indices T, M and S refer to the 200 mb level, the 600 mb level and the surface, respectively.

As a further basic variable the model also includes the surface temperature. Due to the large difference in the effective heat capacity of land and ocean surfaces, it seems necessary to distinguish between the land surface temperature T_L and the ocean surface temperature T_W in order to model the seasonal cycle. The zonally averaged surface temperature T_S may then be written

$$T_S = fT_W + (1 - f)T_L, \qquad [4]$$

where f is the fraction of the latitude circle covered by ocean. Analogous expressions are used for F_S and H_S. The prognostic equation of the land surface temperature is

$$C_L \frac{\partial T_L}{\partial t} = A_L - F_L - H_L \qquad [5]$$

where C_L is the heat capacity of the land surface, A_L the solar radiation absorbed by the surface, F_L the net upward long-wave radiation and H_L the flux of sensible and latent heat from the surface to the atmosphere. Using the same symbols as in eq.[5] but with the index W, instead of L, indicating the ocean surface, the prognostic equation for the temperature T_W is written

$$C_W \frac{\partial T_W}{\partial t} = A_W - F_W - H_W + G \qquad [6]$$

where the last term describes the heating due to the heat transport in the oceans. The magnitude of the heat capacity C_L varies strongly with the physical properties of the surface, but in all cases the value of C_L is more than one order of magnitude smaller than the heat capacity C_W of the upper mixed layer of the ocean. In the numerical experiments, C_L has been treated as a constant with a value of 50 W m^{-2} day K^{-1}, and for C_W the value 3000 W m^{-2} day K^{-1} has been used, corresponding to the heat capacity of a column of sea water 62 m deep.

The heat flux H_S from the surface to the atmosphere is composed of the flux of sensible heat and the flux of latent heat. The flux of sensible heat is computed from the simple linear expression $a(T_S - T_3 - \gamma_S)$ with the value $a = 6$ W m^{-2} K^{-1} and $\gamma_S = 8$ K. The heat flux connected with the evaporation is computed as $b q_S$, where q_S is the saturation mixing ratio at the surface temperature and where b is 4000 W m^{-2} for land surfaces and 8000 W m^{-2} for ocean surfaces. The energy transfer from the surface to the lower layer of the atmosphere is linked to a further small-scale vertical transport, H_M, from the lower to the upper layer. It is assumed, that the contribution from the sensible heat flux to H_M is given as $a(T_3 - T_1 - \gamma_M)$ with $\gamma_M = 24K$. Furthermore it is assumed, that a flux of latent heat contributes to H_M with a value given as a certain fraction (1/5) of the flux bq_S at the surface.

The radiation calculations determining the absorbed solar radiation A and the net flux of long-wave radiation F are described in detail in the paper by Eliasen and Laursen (4) and only a few specific points will be mentioned here. The incident solar flux at the top of the atmosphere as a function of latitude and calendar date is computed from the astronomical parameters as mean values for each day by the procedure given by Berger (6). The distribution of water vapour used by the radiation calculations is expressed by the mixing ratio $q(p) = q_S(p/p_S)^m$ where m as function of latitude and time is determined from the condition that $q = 2.5 \cdot 10^{-6}$, when $p \leqslant 100$ mb. The vertical distribution of the temperature is determined from T_S, T_3 and T_1 using the reasonable approximation that T^4 varies linearly with p and with the further condition that the temperature above the 100 mb level is a constant equal to 210 K. Concerning the clouds it is assumed that the cloud cover is 0.5, independent of latitude and time. The pressure at the mean height of the cloud top and of the cloud bottom is taken as 500 mb and 700 mb, respectively. The albedo at the top of the cloud cover is $\alpha = 0.34 + 0.008(Z-45)$, where Z is the effective mean zenith angle, and the albedo of the cloud base is 0.45. The albedo of the earth's surface is a very important factor in climate modelling, especially due to the ice-albedo feedback. In the present model formulation the following parameterization of the albedo of land and ocean surfaces is applied

$$\alpha_L = \begin{cases} 0.16 & T_L \geqslant 273 \text{ K} \\ 0.16 + 0.06(273 - T_L) & T_L < 273 \text{ K} \end{cases} \quad [7]$$

$$\alpha_W = \begin{cases} 0.07 + 10^{-7}(Z-45)^4 & T_W \geqslant 271 \text{ K} \\ 0.07 + 0.09(271 - T_W) & T_W < 271 \text{ K} \end{cases} \quad [8]$$

with the values restricted to be less than 0.75.

In the paper by Eliasen and Laursen (4) the heat transport in the ocean is described as a horizontal diffusion. In the present variant of the model, however, the ocean heating term G is expressed as a form of Newtonian cooling given by

$$G = K(\tilde{T}_H - T_H) \quad [9]$$

where $T_H = \begin{cases} T_W & T_W \geq 271 \text{ K} \\ 271 \text{ K} & T_W < 271 \end{cases}$ [10]

and $\tilde{T}_H = [f\, T_H][f]^{-1}$ [11]

This treatment of the effect of the heat transport in the oceans seems more in accordance with the modelling of the large-scale heat transport in the atmosphere than a diffusion and in any case it is somewhat more simple. A reasonable ocean heating is obtained by using the constant value $K = 2.0$ W m^{-2} K^{-1}.

When the temperature T_W falls below 271 K, sea ice is formed, and instead of eq.[6], the temperature T_W of the sea ice surface is determined by

$$\rho_I C_I I \frac{\partial T_W}{\partial t} = A_W - F_W - H_W + F_I \quad [12]$$

where I is the sea ice thickness and ρ_I and C_I the density and heat capacity of sea ice, respectively. F_I denotes the upward heat flux through the ice. With a constant value (271 K) of the temperature in the ocean layer below the ice the energy equation for this layer becomes

$$\rho_I L \frac{\partial I}{\partial t} = F_I - G \quad [13]$$

where L is the latent heat of fusion. Instead of using directly an expression for the heat flux F_I, it is assumed that the sea ice thickness may be parameterized as a function of the surface temperature T_W. With this assumption eqs.[12] and [13] together may be written on the same form as [6], but with

$$C_W = \rho_I (C_I I - L \frac{dI}{dT_W}) \quad [14]$$

It seems reasonable to expect that $(-\frac{dI}{dT_W})$ has a maximum value at $T_W = 271$ K and as a simple possibility, the following expression for C_W has been used

$$C_W = \begin{cases} C_O & T_W > T_I \\ C_O - \frac{T_I - T_W}{T_I - T_N}(C_O - C_N) & T_N \leq T_W \leq T_I \\ C_N & T_W < T_N \end{cases} \quad [15]$$

with $C_O = 3000$ W m^{-2} day K^{-1}, $C_N = 200$ W m^{-2} day K^{-1}, $T_I = 271$ K, and $T_N = 261$ K.

Another important factor in the climate system is the effect of snow accumulation. Without having a closed hydrologic cycle formulated in the model, it seems most fair to assume that snow falls, when the surface temperature is below 273° with a rate, prescribed as function of latitude from observed mean values. The snow budget is then computed as described in the paper by Birchfield et al. (7), from where also the used values of snowfall rate are taken.

SENSITIVITY TO THE ORBITAL PARAMETERS

With present-day values of the orbital parameters the model described in the foregoing is able to give a fair simulation of the observed seasonal and meridional temperature variations at the surface as well as at the two tropospheric levels. Also concerning radiative fluxes and planetary albedo the model gives results in good agreement with observed values, cf. Eliasen and Laursen (4). Numerical integrations of the model with different values of the solar constant indicate that the model has a quite significant sensitivity. A decrease of the solar constant by 1% results in a decrease of the mean annual global temperature by 3.1 K. The decrease of the annual mean temperature has a maximum value of 12.5 K at 60°N and another maximum of 10.0 K at 60°S. From 30°S to 30°N the decrease is about 1 K. The corresponding increase of temperature connected with an increase of the solar constant by 1% is essentially more modest with an increase of the mean annual global temperature by 0.9 K, resulting mainly from a temperature increase in the polar region of the northern hemisphere.

The use of climate models for testing the astronomical theory of the ice ages has been discussed recently by Held (8). Also for the present model the sensitivity to variations in the orbital parameters has been studied. The incident solar flux in the model is computed directly from the three parameters : the eccentricity, the obliquity, and the longitude of the perihelion. Steady-state solutions for the model have been obtained for a number of combinations of the possible values of the orbital parameters, and it is found that the most significant decrease of the temperature occurs for the minimum value of the obliquity, namely 22°. Also the variations in the longitude of the perihelion is essential, especially for high value of the eccentricity, of course. The following table illustrates the decrease of temperature from present-day values obtained with the obliquity equal to 22°, the eccentricity equal to 0.05 and with four different values of the longitude of perihelion (December perihelion : $\omega = 90°$).

ω	Δ [T]	ΔT(0°)	ΔT(60°N)	ΔT(60°S)	ΔT(60°N)JULY
0	1.3	0.1	6.5	2.2	10.2
90	2.8	1.2	10.9	10.1	16.7
180	0.2	-0.3	1.1	1.1	- 0.4
270	0.2	-0.2	1.4	1.1	- 0.5

The first column shows the decrease of mean annual global temperature, the next three the decrease of annual mean temperature at the equator, at 60°N and at 60°S, and finally the last column shows the decrease of the temperature in July at 60°N. It is seen that especially for ω=90 the model has a climate essentially colder than today. The decrease of the annual mean surface temperature has a maximum around 60° at both hemisphere and with the largest decrease in the late summer. The line of permanent snow ($T_L < 273$ K) has been displaced southward to 53° at the northern hemisphere. These temperature conditions are quite close to ice age conditions, and back in time combinations of the values of the orbital parameters close to that used above have occured several times, for example at about 116 000 and 231 000 years BP.

REFERENCES

1. Budyko, M.J. : 1969, Tellus 21, pp. 611-619.
2. Sellers, W.D. : 1969, J. Appl. Met. 8, pp. 392-400.
3. Suarez, M.J., and Held, I.M. : 1979, J. Geophys. Res. 84, pp. 4825-4836.
4. Eliasen, E., and Laursen, L. : 1982, Tellus 34, pp. 514-525.
5. Eliasen, E. : 1982, Tellus 34, pp. 228-244.
6. Berger, A.L. : 1978, J. Atmos. Sci. 35, pp. 2362-2367.
7. Birchfield, G.E., Weertman, J., and Lunde, A.T. : 1982, J. Atmos. Sci. 39, pp. 71-87.
8. Held, I.M. : 1982, Icarus 50, pp. 449-461.

PRELIMINARY RESULTS ON THE SIMULATION OF CLIMATE DURING THE LAST DEGLACIATION WITH A THERMODYNAMIC MODEL

J. Adem[1], A. Berger[2], Ph. Gaspar[2], P. Pestiaux[2] and J.P. van Ypersele[2]

[1] Centro de Ciencias de la Atmosfera, UNAM, 04510 Mexico D.F.
[2] Université Catholique de Louvain, Institut d'Astronomie et de Géophysique G. Lemaître, B-1348 Louvain-la-Neuve, Belgium

ABSTRACT

   The objective of the present work is to simulate the equilibrium climate at 5 different stages of the last deglaciation, in order to assess the respective role of different forcings: insolation, ice boundaries and sea surface temperature. We use as forcing the radiation data from Berger (1), the ice sheet boundaries from Denton (2) and the sea surface temperature from CLIMAP (3).

   In these experiments we use Adem's thermodynamic model which is a hemispheric grid model with a realistic distribution of continents and oceans and which includes these three forcings as input data. The procedure used is to simulate first the climate for present conditions and then for the 5 stages of the deglaciation for which Denton gives ice boundaries : 18, 13, 10, 8 and 7 kyr BP.

INTRODUCTION

   The model used in these experiments includes an atmospheric layer of about 10 km thickness and a continental layer of negligible depth. The basic equation is that for the conservation of thermal energy, which when applied to the atmospheric layer and to the ocean (or continent) layer yields two equations that contain the mean tropospheric temperature, the surface temperature and the heating and transport terms as variables. The other conservation laws are used together with semi-empirical relations to parameterize the heating and transport components. These parameterizations provide additional equations which are combined with the thermal energy equations to yield a system of simultaneous equations which are solved in the way described by Adem (4). The integration is performed over the Northern Hemisphere with the use of the basic NMC grid abridged to 512 grid points.

The model has been applied mainly to simulation of the present climate as well as to monthly climate predictions (5,6,7,8,9) and to ocean temperature predictions (10,11). Recently it has been adapted to study the climates of the past. In one paper the seasonal cycle of the climate of 18 kyr BP was simulated (12) and in another one (13) the sensitivity of the ice cap to changes of insolation during the last deglaciation was studied. In the later experiments starting at 18 kyr BP the rate of ablation, due to the changes in insolation, was shown to be small up to 8 kyr BP. In these experiments the ocean temperatures were kept fixed and equal to those at 18 kyr BP.

In the present work preliminary computations are carried out to simulate the equilibrium climate at 5 different stages of the last deglaciation in order to assess the respective role of insolation, ice boundaries and sea surface temperature.

THE NUMERICAL EXPERIMENTS

We use the same seasonal model as in Adem (13), applied to the summer season, except that in the present experiments the ice boundary is prescribed and kept fixed for each of the experiments.

As in previous applications of the model we first compute the present climate and then the climate of the past. The computations are carried out only for the summer season (taken as the average of June, July and August). The present summer climate (normal case) is computed by prescribing the summer normal insolation, surface albedo and sea surface temperature, obtained respectively from Berger (1), Posey and Clapp (14) and U.S. Navy Hydrographic Office (15).

For the past climates, the past insolation data (1) are used. The surface albedo is estimated using Denton's ice sheet extents (2). We assign a value of 70% to the ice sheet surface albedo, except near the pole where we use a value of 80%. Elsewhere we use the normal summer albedo value given by Posey and Clapp (14). We assume that the ocean temperatures vary linearly from 18 kyr BP, for which the CLIMAP values (3) were used, to 10 kyr BP and that from 10 kyr BP to present they are equal to the present temperatures.

The results for 18, 13, 10, 8 and 7 kyr BP are given. Figure 1 shows the average over the whole region of integration of the computed surface (ground) temperature anomaly. The abscissa is the time in kyr BP and the ordinate the ground temperature anomaly in K for different forcings: the thin dotted line (labeled I) is the case when the forcing is only the anomaly of insolation; the dashed line (labeled α), the case when the forcing is the albedo anomaly only; the continuous line ($I+\alpha$) is the case when both insolation and albedo anomalies are included, and the thick dotted line ($I+\alpha+T_s$) is the case when besides the anomalies of insolation and surface albedo, those of ocean temperature are included. The complete solution corresponds to this case.

Figure 2 is similar to Figure 1 but for the mean tropospheric temperature anomaly.

Figure 3 shows the zonally averaged values of the surface (ground) temperature. Parts A, B, C, D correspond to the solutions for 18, 13, 10 and 7 kyr BP. The lines are defined and labelled in the same way as those of Figure 1. For 10, 8 and 7 kyr BP the thick dotted line is not shown because it coincides with the continuous line ($I+\alpha$). Figure 3A shows that for 18 kyr BP the effect of the forcing due to the insolation anomaly is small when applied alone, but increases when applied together with the albedo anomaly, to become non-negligible. Furthermore in this case the ocean temperature has a very important effect.

Figure 4 shows the computed mean tropospheric temperature anomalies in °C. Parts A, B, C and D correspond to 18, 13, 10 and 7 kyr BP for the case when the three forcings are included and the ocean temperature is used as described above. The solution for 8 kyr BP is not shown because it is very similar to that for 10 kyr BP.

REMARKS AND CONCLUSIONS

It follows from Figures 1 and 2 that in the Northern Hemisphere, the insolation effect increases the temperature by about 1°C for 13, 10, and 8 kyr BP and about 0.5°C for 7 kyr BP. However, Figure 4 shows that the effect is much more important in the continental areas reaching values as large as 2.5°C in Asia and 3°C in North Africa.

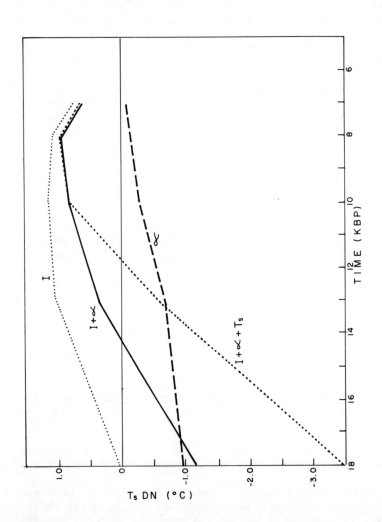

Figure 1 The average over the Northern Hemisphere of the computed ground temperature anomalies for 18, 13, 10, 8 and 6 kyr BP for different forcings. Insolation anomalies alone : thin dotted line (I) ; surface albedo anomalies alone : dashed line (α) ; insolation and surface albedo anomalies : continuous line (I+α) ; insolation, surface albedo and sea surface temperature anomalies : thick dotted line (I+α+T$_S$).

SIMULATION OF CLIMATE DURING THE LAST DEGLACIATION

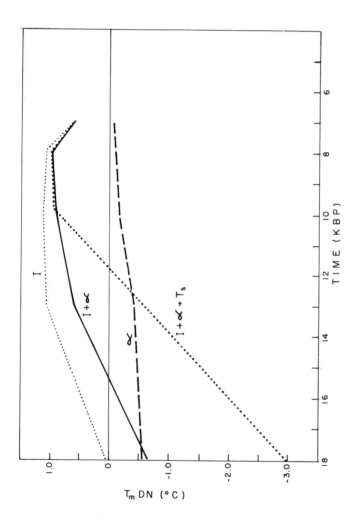

Figure 2 The average over the Northern Hemisphere of the computed mean tropospheric temperature anomalies 18, 13, 10, 8 and 7 kyr BP for different forcings. Insolation anomalies alone : thin dotted line (I) ; surface albedo anomalies alone : dashed line (α) ; insolation and surface albedo anomalies : continuous line (I+α) ; insolation, surface albedo and sea surface temperature anomalies : thick dotted line (I+α+T_S).

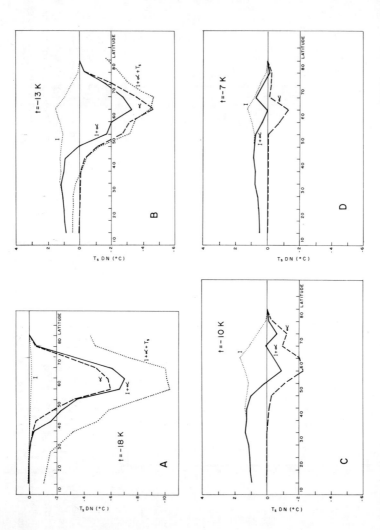

Figure 3 Zonally averaged computed ground temperature anomalies for different forcings. Insolation anomalies alone : thin line (I); surface albedo anomalies alone: dashed line (α); insolation and surface albedo anomalies: continuous line (I+α); and insolation, surface albedo and sea surface temperature anomalies: thick dotted line (I+α+T_s). A,B,C and D for 18, 13, 10 and 7 kyr BP respectively.

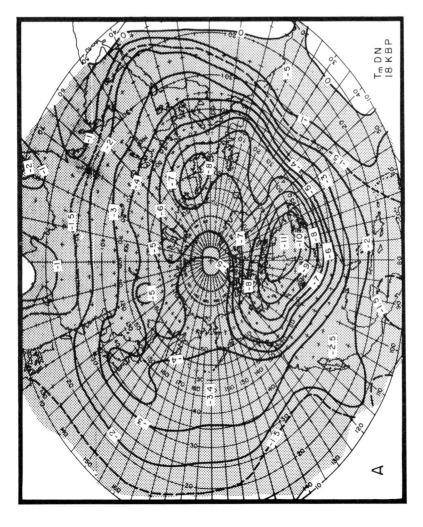

Figure 4A The computed mean tropospheric temperature anomalies for 18 kyr BP.

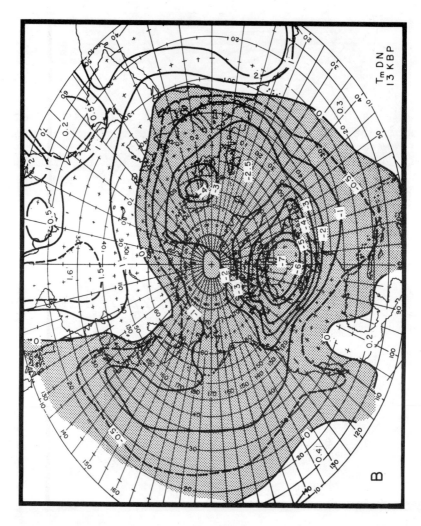

Figure 4B Same as Figure 4A but for 13 kyr BP.

SIMULATION OF CLIMATE DURING THE LAST DEGLACIATION

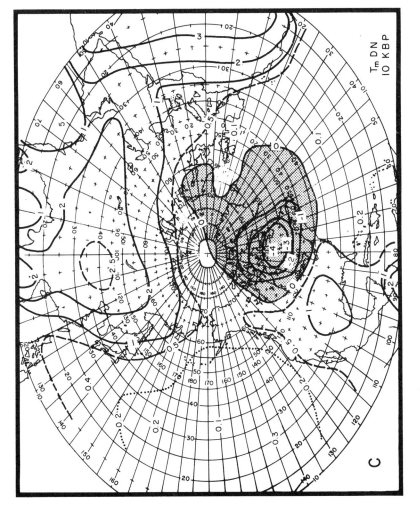

Figure 4C Same as Figure 4A but for 10 kyr BP.

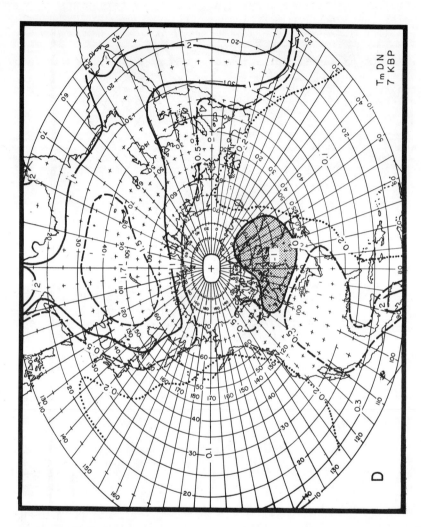

Figure 4D Same as Figure 4A but for 7 kyr BP.

ACKNOWLEDGEMENTS

We are indebted to Jorge Zintzun for his help with the programming and the numerical computations and to José Lauro Ramirez for his assistance in the preparation of the figures.

Ph. Gaspar is supported by the National Fund for Scientific Research (Belgium) ; C. Tricot and J.P. van Ypersele are supported by the Commission of the European Communities (contract n° CL-026-B(D)). These supports are gratefully acknowledged.

REFERENCES

1. Berger, A. : 1978, J. Atmos. Sci. 35, 12, pp. 2362-2367.
2. Denton, G.H., and Hughes, T.J. : 1981, "The Last Great Ice Sheets", J. Wiley, New York.
3. CLIMAP : 1976, Science 191, pp. 1131-1137.
4. Adem, J. : 1979, Dyn. Atmos. Oceans 3, pp. 433-451.
5. Adem, J. : 1964, Mon. Wea. Rev. 92, pp. 91-103.
6. Adem, J. : 1964, Geofis. Intern. 4, pp. 3-32.
7. Adem, J. : 1965, Mon. Wea. Rev. 93, pp. 495-503.
8. Adem, J. : 1970, Mon. Wea. Rev. 98, pp. 776-786.
9. Adem, J., Donn, W.L., and Goldberg, R. : 1982, "Monthly climate predictions for 1980-81 with a thermodynamic model incorporating parameterized dynamics", Proceedings of the Sixth Annual Climate Diagnostics Workshop, pp. 300-310.
10. Adem, J. : 1970, Tellus 22, pp. 410-430.
11. Adem, J. : 1975, Tellus 27, pp. 541-551.
12. Adem, J. : 1981, J. Geophys. Res. 86, pp. 12015-12034.
13. Adem, J. : 1981, Climatic Change 3, pp. 155-171.
14. Posey, J.W., and Clapp, P.F. : 1964, Geofis. Intern. 4, pp. 38-48.
15. U.S. Navy Hydrographic Office 1944, "World Atlas of Sea Surface Temperature", H.O. 228 Washington, D.C.

PART III

MODELLING LONG-TERM CLIMATIC VARIATIONS IN RESPONSE TO ASTRONOMICAL FORCING

SECTION 2 - MODELS WITH AN ICE SHEET

SOME ICE-AGE ASPECTS OF A CALVING ICE-SHEET MODEL

D. Pollard[1]

Climatic Research Institute
Oregon State University
Corvallis, Oregon 97331

ABSTRACT

Further results and extensions of a simple northern hemispheric ice-sheet model are described for the Quaternary ice ages. The model predicts ice thickness and bedrock deformation in a north-south cross section, with a prescribed snow-budget distribution shifted vertically to represent the orbital perturbations. An ice calving parameterization is added, crudely representing proglacial lakes or marine incursions that attack the ice whenever the tip drops below sea level. With this extension the model produces a large 100 kyr response with complete deglaciations, and agrees fairly well with the $\delta^{18}O$ deep-sea core records.

The observed phase-correlation between the 100 kyr cycles and eccentricity is examined. First, the model is shown to give a $\sim$ 100 kyr response to nearly any kind of higher-frequency forcing. Although over the last two million years the model phase is mainly controlled by the precessional modulation due to eccentricity, over just the last 600 kyr the observed phase can also be simulated with eccentricity held constant. A definitive conclusion on the phase-control of the Quaternary ice ages is prevented by uncertainty in the deep-sea-core time scales before $\sim$600 kyr BP

As an alternative to the calving mechanism, a generalized parameterization for the feedback effects of a North Atlantic deglacial meltwater layer is examined. Much the same ice-age simulations and agreement with the $\delta^{18}O$ records as with the calving mechanism can still be obtained, but only with implausibly sensitive feedbacks.

Alternate treatments of some other model components are described briefly, including the coupling to a global seasonal climate model. Despite these modifications, nearly the same ice-age simulations as with the basic model can still be obtained.

INTRODUCTION

One of the main tasks in ice-age modelling is the simulation of the fluctuations of northern hemispheric ice sheets over the last million years. Oxygen isotope measurements in deep-sea cores are generally believed to be good proxy indicators of northern hemispheric ice-sheet volume, although they may also be influenced by other variables such as temperature and Antartic ice-sheet size (1,2,3). The dominant features in these records are "glacial" cycles with an average period of about 100 kyr and with full glacial to interglacial amplitudes. Many cycles are terminated by a relatively rapid retreat from maximum ice-sheet size to conditions like those at present, as occured for instance between about 18 and 6 kyr BP The records also show minor oscillations of smaller magnitude with spectral peaks at roughly 20 and 40 kyr, and these have been well correlated with insolation variations due to the orbital perturbations of about the same periods, in accordance with the "Milankovitch" theory (4).

Several simple climate and/or ice-sheet models have been able to simulate the higher-frequency oscillations as a direct response to the orbital forcing (5,6,7,8,9). Almost all of the spectral power of the orbital forcing occurs at periods around 22 and 41 kyr due to variations in the precession and obliquity, respectively, and these models for the most part responded linearly with amplitudes similar to or greater than the observed minor oscillations. However, little or no sign of the dominant 100 kyr glacial cycles was produced.

Recently Oerlemans (10) and Birchfield et al. (11), using simple ice-sheet models with orbital forcing, obtained results that agreed with some aspects of the observed glacial cycles. This was achieved by including a realistic time lag in the bedrock deformation under the ice load, which caused some of the ice-sheet retreats to be amplified. In the model of Birchfield et al. (11), these retreats coincided well with the observed deglaciations, but some unrealistic features remained : there was still too little 100 kyr spectral power compared to the higher frequency peaks, and in some of the major retreats their model ice sheet did not vanish completely ^{2}.

These shortcomings were eliminated by adding to the model a crude parameterization of calving by proglacial lakes and/or marine incursions (12,13,14). Calving in the model occurs at the southern ice-sheet tip only during the rapid retreats at roughly 100 kyr intervals, further accelerating the retreats to achieve more complete deglaciations. The resulting simulations then agreed well with the deep-sea-core records and their power spectra, especially over the last 400 kyr.

In the present paper the ice-sheet model with calving is described and the results are analyzed further. First the model formulation is outlined in section 2, and the basic results are presented. In section 3 the relationships between the model's 100 kyr response, the orbital forcing and the variations of eccentricity are examined more closely. Several investigators have noted the agreement between the phases of eccentricity and the observed glacial cycles over the last 600 kyr (41), although the magnitudes of the individual cycles are uncorrelated (15,16). Variations in eccentricity serve mainly to modulate the ~22 kyr precessional forcing, and cause only a very small amount of 100 kyr power in the insolation variations. Consequently some highly selective response would presumably be required to maintain a phase-lock with eccentricity. These issues are investigated here by running the model with a variety of imposed forcing functions to identify which components of the orbital forcing are controlling the phase of the model's 100 kyr response.

In section 4 another mechanism is examined that replaces the role of calving, involving the feedback effects of ice-sheet runoff into the North Atlantic during deglaciations (17,18). In section 5 several alternate treatments of various model components are described briefly, including the coupling to a global seasonal climate model. In section 6 the results and conclusions are summarized.

It should be noted that other mechanisms might be important in the 100 kyr glacial cycles. These include long-term free internal oscillations involving the deep ocean (19,20), and "tunneling" due to stochastic forcing (21,22). Both of these are absent in the present model.

MODEL FORMULATION AND BASIC RESULTS

Only a terse formulation is given here; the model is described more fully in Pollard (12). Much of the model is similar to those of Oerlemans (10) and Birchfield et al. (11). The overall structure of the model is shown in Figure 1.

Ice thickness, h, and the elevation of the bedrock surface, h', are predicted as functions of latitude, representing a

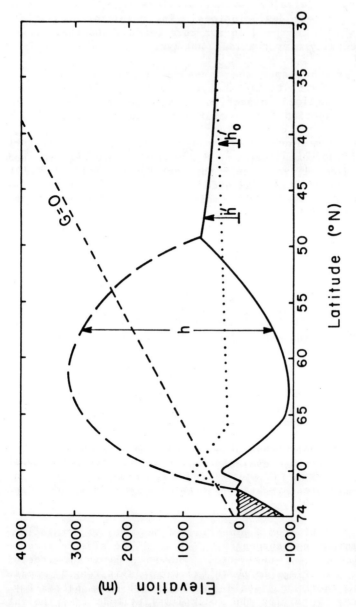

Figure 1 Schematic structure of the model. Here h is the ice-sheet thickness, h' is the bedrock elevation above present mean sea level (full line), h_0' is the equilibrium crustal topography (dotted line), and G=0 is the equilibrium line of the prescribed snow budget (short-dashed line). A northern body of water is shaded; during deglaciations a similar water body can also form and cause calving at the southern edge of the ice sheet when the tip drops below sea level.

north-south profile along a typical flow line through the Laurentide or Scandinavian ice sheets.

$$\frac{\partial h}{\partial t} = A \frac{\partial}{\partial x} [h^\alpha \left|\frac{\partial(h+h')}{\partial x}\right|^\beta \frac{\partial(h+h')}{\partial x}] + G(h+h',x,orbit) \qquad [1]$$

$$\frac{\partial h'}{\partial t} = \nu \frac{\partial}{\partial x^2} [h' - h'_0(x) + rh] \qquad [2]$$

Here t is time and x is distance to the south. Eq. [1] uses a vertically integrated approximate ice flow law with east-west flow neglected (11). The ice flow coefficient $A = 5.77 \times 10^{-4}$ m^{-3} yr^{-1}, $\alpha=5$ and $\beta=2$. G is the net annual mass balance on the ice surface, representing the current distributions of snowfall and ice melt; its dependence on the surface elevation h + h' follows Oerlemans (10):

$$G = \begin{cases} a(h+h'-E) - b(h+h'-E)^2 & \text{if } h+h'-E \le 1500m \\ 0.56 & \text{if } h+h'-E > 1500m \end{cases} \quad m\ yr^{-1} \qquad [3]$$

where $a = 0.81 \times 10^{-3}$ yr^{-1}, $b = 0.30 \times 10^{-6}$ m^{-1} yr^{-1}; all elevations are referred to the present mean sea level. E is the equilibrium-line altitude at which the mass balance is zero, and is taken to have a constant slope in latitude. It is shifted uniformly in response to the orbital perturbations according to

$$E = E_0 + s(x - x_0) + k\Delta Q \qquad [4]$$

where ΔQ is the difference in the summer half-year insolation from that of the present at the arbitrarily selected latitude of 55°N, calculated from the current orbital parameters (23). Unless otherwise stated, $k=35.1$ m per $(W\ m^{-2})$, $E_0=550$ m, $s=10^{-3}$ and x_0 corresponds to 70°N.

Eq. [2] follows from a linearized flow model in a relatively thin asthenospheric channel (20,24), with the elastic response of the lithosphere neglected. However, permanent crustal topography is prescribed in the $h'_0(x)$ term. This roughly represents a North American profile running south from Baffin Bay, piecewise-linearly joining the following [latitude (°N), height (m)] points : (74,-500), (70,850), (66,200), (40, 400), (30, 400). In eq. [2] the asthenospheric flow coefficient $\nu = 100$ $km^2 yr^{-1}$, and r=0.3 is the ratio of ice density ρ_1 to rock density ρ_A.

Ice calving is added as described in Pollard (12). This crudely represents accelerated flow and wastage at the southern tip of the ice sheet due to proglacial lakes and/or marine incursions. These processes have been emphasized by Andrews (25) and Denton and Hughes (26) as possibly being important during the last deglaciation. Here the ice budget G in [3] is set equal to a large negative value at ice-sheet points between the tip and the first interior ice-sheet point that cannot be floated by the hydrostatic head of the water :

$$G(x_{i+1}) = -20 \text{ m yr}^{-1} \quad \text{if} \quad \begin{cases} \rho_I h(x_i) < \rho_W (S - h'(x_i)) \\ \text{and } h'(x_{i+1}) < S \end{cases} \quad [5]$$

where ρ_W is the density of the water, S is the current sea level (taken to be zero), and i is the spatial grid index increasing into the ice sheet. In practice the southern tip is only attacked in this way during deglaciations and only the outermost 50 to 150 km of ice sheet is affected. The calving mechanism also attacks the northern tip and maintains it at $\sim 72°N$, but does not penetrate farther into the ice sheet because of the high mountain centered on 70°N.

Eqs. [1] and [2] are numerically integrated forward in time using implicit Newton-Raphson estimations for all terms except those in G. Time steps of 50-100 years are used with a latitudinal resolution of 55.5 km (=0.5°). Latitudinal boundary conditions are taken to be that $h' = h'_0$ at the model extremities of 74°N and 30°N.

Three ice-age runs are shown in Figure 2 with progressively more complex model versions. Each is compared with an oxygen-isotope record. In Figure 2a the bedrock depression is always in isostatic equilibrium with the ice load [$\nu = \infty$ in [2]], and calving is supressed. This is similar to results of earlier models (5), with the ice sheet responding directly to the orbital forcing but with no trace of a 100 kyr cycle. In particular the observed rapid deglacial retreats are lacking, for instance the last deglaciation from ~ 18 to 6 kyr BP.

Figure 2b is produced with $\nu = 100 \text{ km}^2 \text{yr}^{-1}$, corresponding to a bedrock lag time of ~ 10 kyr for the broadest features (24). Here, as in Oerlemans (10) and Birchfield et al. (11), the bedrock lag in [2] acts to amplify major retreats (for instance around 10, 130, 220 and 340 kyr BP) by the following mechanism : at a glacial maximum the bedrock depression is relatively deep, so that any subsequent retreat is accelerated by the increased ice melt at the lower elevations near the southern tip. In Figure 2b, however, this amplifying loop is usually not powerful enough to produce complete deglaciations. Some complete degla-

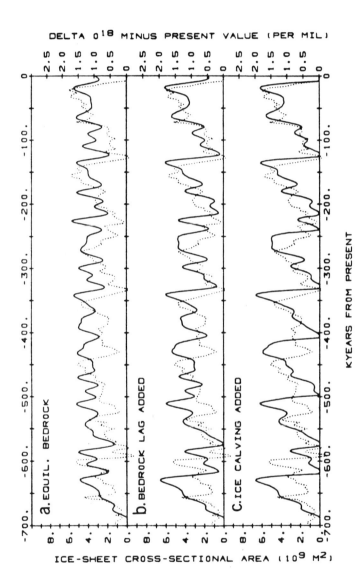

Figure 2 Ice-age simulations, i.e., total cross-sectional area of ice in the (latitude, height) plane versus time. The dotted curve in each panel is an oxygen-isotope deep-sea core record, relative to its most recent $\delta^{18}O$ value, redrawn from Emiliani (43, Fig. 2).
(a) With equilibrium bedrock ($\nu_2 = \infty$) and without calving.
(b) With bedrock lag ($\nu = 100 \text{ km}^2\text{yr}^{-1}$) and without calving.
(c) With bedrock lag ($\nu = 100 \text{ km}^2\text{yr}^{-1}$) and with calving (Eq. 5).

ciations can be obtained by increasing the orbital sensitivity, k, but then the amplitude of the direct 20 and 40 kyr response becomes too great (12).

The calving mechanism [5] is included in Figure 2c. Calving sets in at the southern ice-sheet tip at about the mid-point of each deglaciation, and enables the ice sheet to vanish completely without resorting to an overly large value of k. The overall agreement with the deep-sea-core record is improved, especially after about 400 kyr BP. Before that time, the agreement with the phase of the main glacial cycles is apparently lost but the agreement before 400 kyr BP is considerably better with the isotopic time scale presented elsewhere in this volume by Imbrie et al. (44) ; this point is discussed further in next Section. The ice thicknesses and latitudinal extents for a part of this run are shown in Figure 3.

Figure 4. shows power spectra for the three model curves in Figure 2. A gradual increase in the 100 kyr peak is seen as each new mechanism is added to the model. The spectrum with the calving mechanism has relative peak sizes in good agreement with those observed (15).

ORBITAL CONTROL OF THE PHASE OF THE 100 KYR GLACIAL CYCLES

The phase of the observed 100 kyr glacial cycles correlates well with that of eccentricity over the last 600 kyr, despite the relatively small direct effect of eccentricity on the orbital forcing, as mentioned in section 1. This correlation is shown in Figure 5, where the $\delta^{18}O$ record of the core V28-239 from Shackleton and Opdyke (28) is plotted using a time scale linearly interpolated between the stage-boundary ages given in Pisias and Moore (29, table 1). The correlation is apparently lost before 600 kyr BP, but this could easily be due to uncertainties in dating the cores (Pisias, personal communication). The other $\delta^{18}O$ curve in Figure 5 is from Johnson (16) who revised the time scale slightly by tuning to features in the orbital insolation signal; not surprisingly, this curve agrees in phase with eccentricity back to 800 kyr BP. The magnitudes of the individual glacial cycles are not correlated with those of eccentricity, and this is perhaps partly why cross-spectral analyses between the two signals are not totally convincing (30, 31), showing coherency peaks at 80 to 95 kyr rather than at the dominant 100 to 120 kyr peaks in the power spectra.

Figure 6 shows an extended run of the model with calving. The ~100 kyr correlation with the phase of eccentricity is apparent, with deglaciations and periods of little or no ice coinciding with locally positive departures of eccentricity. In

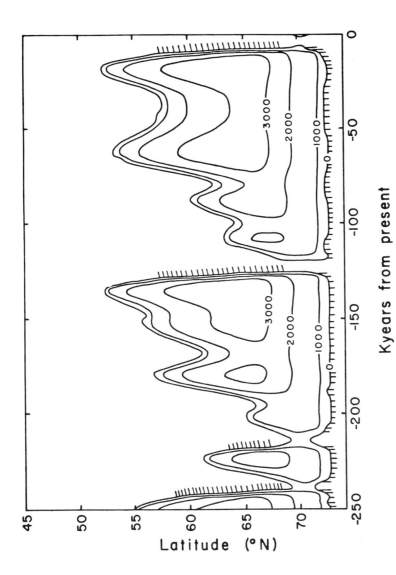

Figure 3 Contours of ice thickness h(m) for the last 250 000 years of the run shown in Figure 2c using the model with calving. Hatching at the ice-sheet edges shows where the tip regions are attacked by calving according to Eq. (5).

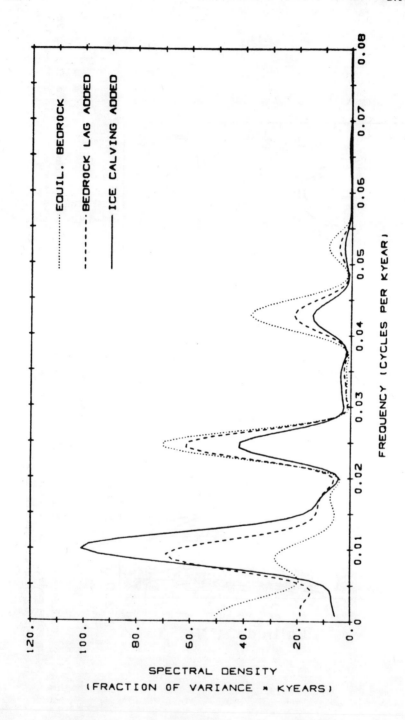

Figure 4 Power spectra for the model simulations of Figure 2, generated via the autocovariances as outlined in Pollard (12).

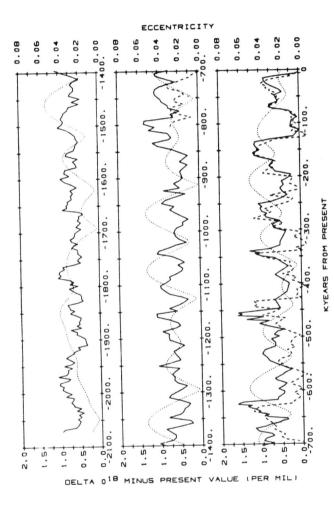

Figure 5 Full curve : oxygen-isotope record of deep-sea core V28-239 from Shackleton and Opdyke (28), relative to its most recent $\delta^{18}O$ value, and with a time scale obtained from Pisias and Moore (29) (see text).
Dashed curve : oxygen-isotope record from Johnson (16, Fig. 11A), relative to its most recent $\delta^{18}O$ value. The $\delta^{18}O$ values are those of V28-238 (42) after 436 000 yr BP, and V28-239 before that date, with slightly revised time scales.
Dotted curve : variations of eccentricity (23).

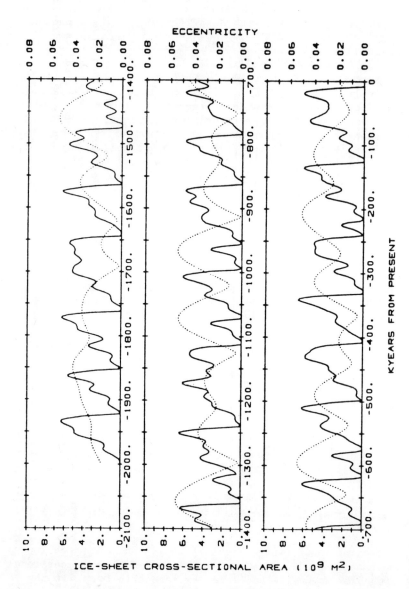

Figure 6 Full curve : ice-age simulation of the calving model. Dotted curve : variations of eccentricity (23).

contrast to the V28-239 record, this correlation is maintained by the model over the last two million years. The model phase does not depend on the choice of initial conditions, since other choices converge onto the same curve within one or two glacial cycles. This correlation and other related issues are examined below by running the model with a variety of forcing functions.

First, Figure 7 shows that the model's $\sim$100 kyr response does not depend on the exact form of the forcing. The model ice sheet has a natural tendency to respond to nearly any kind of forcing by growing slowly to a maximum size, at which point calving and deglaciation can be triggered. (With no forcing the model does not exhibit any internal oscillations.) With sinusoidal forcing as in Figures 7a-e, the period of the main long-term response is always a multiple of the forcing period falling in the 80 to 100 kyr range. The amplitude of the forcing at each frequency has to be adjusted so that calving can be triggered, but the amplitude of the long-term response is still mainly determined by the internal dynamics. A "white-noise" forcing is used in Fig. 7f by replacing $k\Delta Q$ in [4] with a random term, rectangularly distributed between $\pm$ 600 m and changed every 2 kyr. The model's main response is still a 100 kyr cycle as above. [Although not envisioned here as a real stochastic forcing, this type of white noise could possibly be due to internal climatic oscillations of a few thousand years period, such as those modelled by Saltzman (32) and Ghil and Le Treut (20) involving sea ice and/or the deep ocean.]

With the actual orbital forcing, the phase of the $\sim$100 kyr response could conceivably be (i) quasi-random, as found in the model of Oerlemans (10); (ii) controlled by the direct influence of eccentricity on the annual mean insolation; (iii) controlled by the modulation of the amplitude of the precessional cycle due to eccentricity; and (iv) controlled by the combined forcing of obliquity and precession independently of eccentricity (33). According to the present model, case (i) is ruled out since the response is independent of initial conditions and small parameter variations. Case (ii) is also ruled out since in model runs with fixed precession and obliquity, and insolation averaged over the full year in [4], the very small variations in ΔQ [on the order of $(1-ecc.^2)^{1/2}$] yield a negligible model response.

To examine case (iii), Figure 8 shows a run with obliquity held constant, so the only forcing is the $\sim$ 22 kyr precessional cycle with its amplitude modulated by eccentricity. (It should be mentioned that the relative importance of obliquity and precession in simple models is influenced by the choice of latitude, here 55°N, at which the insolation is computed. However, the results presented here still serve to illustrate the various

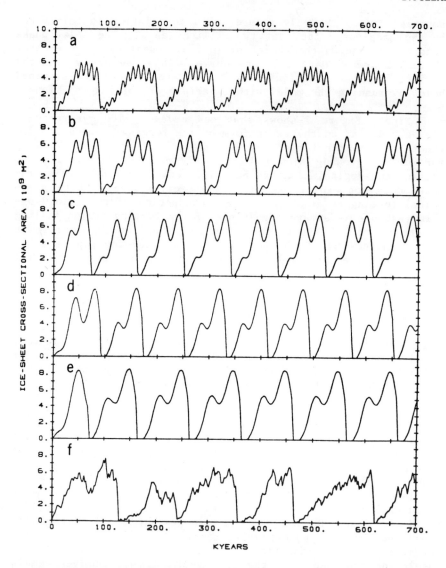

Figure 7 (a)-(e) : Response of the calving model to sinusoidal forcing, in which the $k\Delta Q$ term in (4) is replaced by $H \sin(2\pi t/T)$ where t is time and :
(a) T = 10 000 yrs, H = 800 m.
(b) T = 20 000 yrs, H = 500 m.
(c) T = 30 000 yrs, H = 400m.
(d) T = 40 000 yrs, H = 400m.
(e) T = 50 000 yrs, H = 300m.
 (f) : As above with "white noise" forcing (see text) and with E_0=650 m in (4).

SOME ICE-AGE ASPECTS OF A CALVING ICE-SHEET MODEL

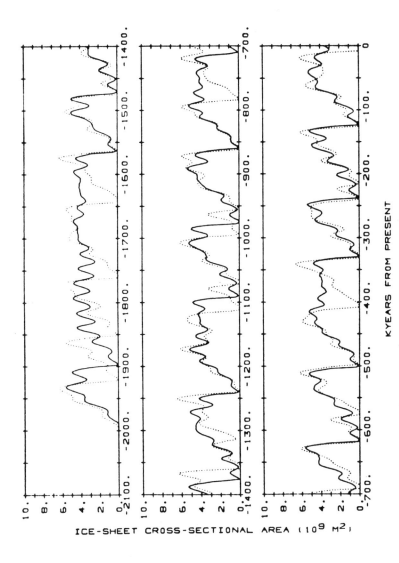

Figure 8 Full curve : ice-age simulation of the calving model but with obliquity held fixed at its present value. Dotted curve : ice-age simulation with full orbital forcing (as in Fig. 6).

possibilities.) Compared to the full orbital case in Figure 6, some of the deglaciations are now missing, and several others are shifted by one precessional cycle, but the overall phase of the long-term cycles is maintained. Taken at face value this implies that the precessional modulation by eccentricity is indeed responsible for the phase of the glacial cycles and the observed correlation with eccentricity.

Figure 9, however, shows a run with eccentricity held fixed so the only forcing is due to obliquity and unmodulated precession [case (iv) above]. Perplexingly, over the last 500 kyr the phase of the ∼100 kyr response again agrees with the full orbital case. Before 600 kyr BP this agreement is generally lost, but unfortunately the uncertainty in the deep-sea-core time scales before that time is too large to determine which model result is most realistic.

We are left with two possibilities :
(i) The real glacial cycles are controlled by eccentricity via the precessional modulation, and with improved dating the $\delta^{18}O$ records would show a phase-lock with eccentricity over the last two million years. In this case Figure 6 is a realistic simulation over that time, and the phase agreement in Figure 9 over the last 500 kyr is coincidental.
(ii) The real glacial cycles are controlled by the combined obliquity and precessional cycles independently of eccentricity. In this case the model result in Figure 6 overestimates the influence of eccentricity, and the phase agreement between the observed glacial cycles and eccentricity over the last 600 kyr is coincidental.

These two possibilities could be distinguished by improved dating of deep-sea cores before 600 kyr BP in order to see if the phase-lock with eccentricity is maintained. In the case of V28-239, this may be hindered by the change in character of the record at about 900 kyr BP, before which the amplitude of the longer-period components are markedly reduced (29).

This change and other long-term variations, such as the 400 kyr signal found in some cores (31), are not simulated in the model run of Figure 6. Suggested causes of these variations are orogeny and erosion by the northern hemispheric ice sheets themselves (29), and Antarctic ice-sheet fluctuations (31). These processes are beyond the scope of the present model, and Figure 6 shows only that the response to the orbital forcing alone has been relatively constant over the last two million years.

SOME ICE-AGE ASPECTS OF A CALVING ICE-SHEET MODEL

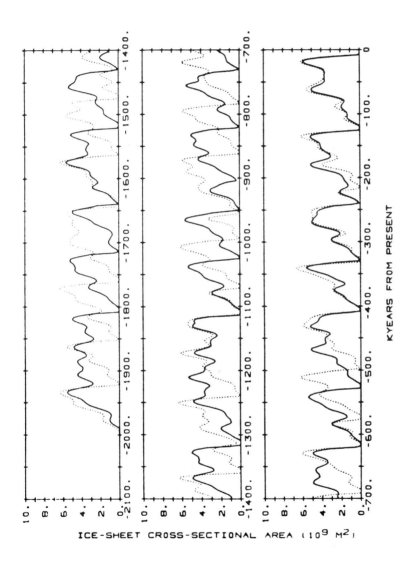

Figure 9 Full curve : ice-age simulation of the calving model but with eccentricity held fixed at 0.02. Dotted curve : ice-age simulation with full orbital forcing (as in Fig. 6).

GENERALIZED MELTWATER PARAMETERIZATION

The main function of the calving mechanism is to accelerate the major retreats so that complete deglaciations can be achieved. Calving is triggered only after a maximum ice-sheet stand has caused a deep bedrock depression, followed by a fairly strong negative orbital anomaly that initiates a retreat of the southern tip into the depression. Alternative mechanisms may exist within the real climate system which are capable of performing the same function, but any candidate mechanism would have to be influenced by the rate of change of ice-sheet size (as opposed to the static influence of the ice sheet such as albedo feedback). For instance, if the meltwater layer from ice-sheet runoff in the North Atlantic could alter the climate so as to produce more ice wastage (17,18), this feedback loop could amplify itself during major retreats to produce complete deglaciations. These authors suggest that the fresh-water input stabilizes and reduces the thickness of the oceanic mixed layer, leading to (i) warmer summers with greater ice-sheet melt, and (ii) colder winters with greater sea-ice extent and less snowfall on the ice sheets.

A thorough investigation of this feedback loop would require a reliable atmospheric and oceanic climate model. Here, the present model is used in a preliminary way, mainly to estimate the strength of the feedback required for good ice-age simulations. The calving mechanism is suppressed and another term is added to [4] to shift the snow-budget pattern vertically, depending on the current net rate of change of ice-sheet volume ($\dot{V}$). (To do this a straightforward iterative procedure is needed to solve for one year's climate.) After some experimenting with the imposed dependence on $\dot{V}$, it was found that relationships like that in Figure 10 are required to produce good ice--age simulations. Here the snow-budget pattern is unaffected above a certain "trigger" value of $\dot{V}$, but below that value (which occurs only during major retreats) it is raised sharply. The resulting ice-age simulation shown in Figure 11 is nearly the same as the basic run with calving in Figure. 2c, with the jump in Figure 10 activated only during deglaciations at about the same times as the calving mechanism was.

The jump in Figure 10 implies a raising of the snow-budget pattern by 500 m [roughly equivalent to a summer temperature increase of 3°C or a winter snowfall reduction of 70% (13)] for a change of $\dot{V}$ corresponding to 8 centimeters of annual meltwater input to the North Atlantic; furthermore, this sensitivity must be somehow suppressed at values of $\dot{V}$ above the trigger value. As noted in Pollard (13), this relationship seems somewhat implausible, but could be tested with more sophisticated climate models. If the jump in Figure 10 is made much less steep, the

SOME ICE-AGE ASPECTS OF A CALVING ICE-SHEET MODEL

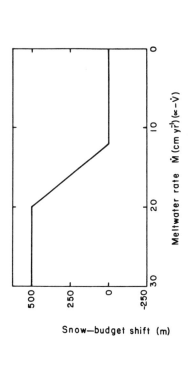

Figure 10 Relationship between the vertical shift of the snow-budget distribution and the net rate of change of ice-sheet size ($\dot{V}$), imposed in Eq. (4) for the generalized meltwater parameterization. $\dot{V}$ is equivalently expressed here as the thickness of the meltwater layer added per year to the North Atlantic (M) given by $-(3 \times 10^{-7} \text{ m}^{-1}) \times d$ (ice-sheet cross-sectional area)/dt.

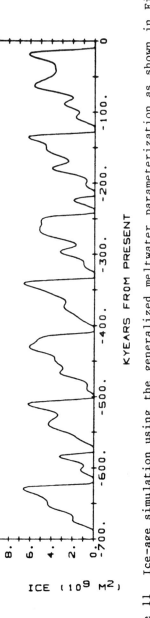

Figure 11 Ice-age simulation using the generalized meltwater parameterization as shown in Figure 10.

feedback loop does not self-amplify when $\dot{V}$ falls below the trigger value in ice-age runs, and most of the retreats are incomplete. These results suggest that mechanisms attacking the ice sheet from "below", i.e. calving or perhaps basal surging, are more promising than those from "above", i.e., reduced snowfall or increased surface melt, in explaining the observed rapid deglaciations.

ALTERNATE TREATMENTS OF MODEL COMPONENTS

The model results above are reasonably stable against small parameter variations (12), and this type of sensitivity is not pursued here. It is felt that more uncertainty stems from the variety of possible representations for some of the model components. Several alternate treatments of the more uncertain model components have been investigated, and are outlined briefly below.

Bedrock response

Although a thin asthenospheric channel was assumed for eq. [2], the profile of viscosity in the mantle is not well known (34,35) and may correspond more closely to an infinitely deep half-space. The latter case has not been tested here since it apparently cannot be reduced to one partial differential equation (36). However, one different type of bedrock response has been tested : a local vertical relaxation with no horizontal communication, as used by Oerlemans (10) and Birchfield et al. (11). With this change, the calving model can be tuned to yield much the same ice-age simulations as above, with nearly as good agreement with the $\delta^{18}O$ records (14).

Ice-shelf processes

The parameterization of calving in eq. [5] is based only on the concept that an unconfined floating ice-sheet tongue will discharge ice much faster than a non-floating tongue. While this appears to be generally borne out by models and observations, a more physical approach would be desirable. Fairly simple models of ice shelves coupled to marine ice sheets have been developed and applied to the Ross ice-shelf sector in West Antarctica (26 ch.7, 27, 37), and to the St. Lawrence sector of the Laurentide ice sheet (38). As described in Pollard (14), a similar ice-shelf representation has been used to replace eq. [5] in the present model, and again much the same ice-age simulations were obtained.

Seasonal climate

One shortcoming of the prescribed snow budget in eqs [3] and [4] is the arbitrary tuning of k, the sensitivity to the orbital perturbations. In addition, any albedo feedback of the ice sheet is absent in the present model. In principle these shortcomings can be eliminated by coupling to a reliable global seasonal climate model of the atmosphere and upper ocean. Unfortunately the climate models used so far in this field have shown large uncertainties in the key sensivities "∂(temperature)/∂(orbital perturbation)" and "∂(temperature)/∂(ice-sheet size)"; these values vary by factors of two or more from model to model (7,8,39). These uncertainties will probably exist until consistent results are obtained with coupled atmospheric and oceanic general circulation models.

Nevertheless, the present ice-sheet model has been coupled to a simple one-level energy-balance climate model with both hemispheric and longitudinal land-ocean contrast (13). Despite some differences in the forms of the orbital sensitivity, much the same ice-age simulations as above were still obtained. These results are not presented here since it is felt that the basic sensitivities of the climate model are no less uncertain than those mentioned above. The realism of the ice-age results with the coupled climate-ice sheet model may reflect not so much the quality of the climate model but rather the robustness of the calving ice-sheet model to details of the orbital forcing.

SUMMARY AND CONCLUSIONS

Some "retreat mechanism" triggered only during deglaciations is required to account for the major ice-sheet retreats observed in the deep-sea core records. With the bedrock lag alone only relatively weak retreats are possible, hindering satisfactory ice-age simulations. The retreat mechanism can be triggered only after the ice sheets have reached maximum size and caused a large bedrock depression, followed by a fairly strong negative orbital anomaly; the retreat mechanism is turned off when the ice sheet has shrunk considerably and/or the bedrock has rebounded sufficiently. With such a mechanism, a simple ice-sheet model can yield fairly realistic records of ice volume and their power spectra over the last several 10^5 years. The calving mechanism in the present model is triggered by the submergence of the southern ice-sheet tip as it descends into the lagged bedrock depression, whereas the generalized meltwater mechanism is triggered by the net surface wastage over the whole ice sheet. Within the limitations of the present model, the calving mechanism seems the more plausible process.

The model responds with a $\sim$100 kyr glacial cycle to nearly any kind of higher-frequency forcing. The amplitude of this cycle is mainly determined by the internal dynamics. Without any external forcing the model does not exhibit internal oscillations, but this depends on the imposed crustal topography and boundary conditions [see appendix of Pollard (14)].

With actual orbital forcing, the phase of the model's ~100 kyr response is controlled mainly by eccentricity via its modulation of the precessional cycle. However, the same phase can be produced by obliquity plus precessional forcing alone with fixed eccentricity, but only over the last 600 kyr. A definite identification of the main phase-control of the observed glacial cycles is prevented by the uncertainty in the deep-sea core time scales before about 600 kyr BP

As cautioned in Pollard (12), the simplicity of the model should be kept in mind when considering these conclusions. Some attempts have been made to increase confidence in the model by using different forms for the more uncertain model components. Much the same results as above have been obtained with these modifications, but further testing would be valuable including (i) an infinitely deep half-space mantle, (ii) a more physical calving and ice-shelf treatment and more explicit models of the proglacial lakes and marine incursions, and (iii) a more reliable global seasonal climate model. In addition, the reduction to one horizontal dimension of the Keewatin-Hudson Bay-Labrador--St. Lawrence complex is particularly drastic, and future model developments should include east-west structure (9).

ACKNOWLEDGMENTS

I am grateful to Robert G. Johnson and Niklas G. Pisias for supplying digital records of the oxygen-isotope curves used in Figure 5. Also I thank Niklas Pisias for a helpful discussion, Michael E. Schlesinger for suggesting the "white-noise" experiment in Figure 7, Susan Hayes for typing the manuscript, and Ron Selig for drafting some of the figures. This research was supported by the Climate Dynamics Section of the NSF under grant ATM-8019762.

REFERENCES

1. Dansgaard, W., and Tauber, H. : 1969, Science 166, pp. 499-502.
2. Chappell, J., and Veeh, H.H. : 1978, Bull. Geol. Soc. Amer. 89, pp. 356-368.
3. Budd, W.F. : 1981, Int. Assoc. Hydrol. Sci. 131, pp. 441-471.

4. Hays, J.D., Imbrie, J., and Shackelton, N.J. : 1976, Science 194, pp. 1121-1132.
5. Weertman, J. : 1976, Nature 261, pp. 17-20.
6. Schneider, S.H., and Thompson, S.L. : 1979, Quaternary Research 12, pp. 188-203.
7. Suarez, M.J., and Held, I.M. : 1979, J. Geophys. Res. 84, pp. 4825-4836.
8. Pollard, D., Ingersoll, A.P., and Lockwood, J.G. : 1980, Tellus 32, pp. 301-319.
9. Budd, W.F., and Smith, I.N. : 1981, Int. Assoc. Hydrol. Sci. 131, pp. 369-409.
10. Oerlemans, J. : 1980, Nature 287, pp. 430-432.
11. Birchfield, G.E., Weertman, J., and Lunde, A.T. : 1981, Quaternary Research 15, pp. 126-142.
12. Pollard, D. : 1982a, Nature 296, pp. 334-338.
13. Pollard, D. : 1982b, "A coupled climate-ice sheet model applied to the Quaternary ice ages.", Climatic Research Institute, Oregon State University, Corvallis, pp. 43. Also J. Geophys. Res. 88, pp. 7705-7718.
14. Pollard, D. : 1982c, "Ice-age simulations with a calving ice-sheet model.", Climatic Research Institute, Oregon State University, Corvallis, in press. Also Quat. Res. 20, pp. 30-48.
15. Imbrie, J., and Imbrie, J.Z. : 1980, Science 207, pp. 943-953.
16. Johnson, R.G. : 1982, Quaternary Research 17, pp. 135-147.
17. Adam, D.P. : 1975, Quaternary Research 5, pp. 161-171.
18. Ruddiman, W.F., and McIntyre, A. : 1981, Science 212, pp. 617-627.
19. Sergin, V. Ya. : 1979, J. Geophys. Res. 84, 3191-3204.
20. Ghil, M., and Le Treut, H. : 1981, J. Geophys. Res. 86, pp. 5262-5270.
21. Benzi, R., Parisi, G., Sutera, A., and Vulpiani, A. : 1982, Tellus 34, pp. 10-16.
22. Nicolis, C. : 1982, Tellus 34, pp. 1-9.
23. Berger, A.L. : 1978, J. Atmos. Sci. 35, pp. 2362-2367.
24. Walcott, R.I. : 1973, Ann. Rev. Earth Planet. Sci. 1, pp. 15-37.
25. Andrews, J.T. : 1973, Arc. and Alp. Res. 5, pp. 185-199.
26. Denton, G.H., and Hughes, T.J., Eds. : 1981, "The last Great Ice Sheets", Wiley, New York, 484 pp.
27. Weertman, J. : 1974, J. Glaciology 13, pp. 3-11.
28. Shackleton, N.J., and Opdyke, N.D. : 1976, Geol. Soc. Amer., Memoir 145, pp. 449-464
29. Pisias, N.G., and Moore, T.C., Jr. : 1981, Earth Planet. Sci. Lett. 52, pp. 450-458.
30. Kominz, M.A., Heath, G.R., Ku, T.L., and Pisias, N.G. : 1979, Earth Planet. Sci. Lett. 45, pp. 394-410.
31. Moore, T.C., Jr., Pisias, N.G., and Dunn, D.A. : 1982 Marine Geol. 46, pp. 217-233.

32. Saltzman, B. : 1982, Tellus 34, pp. 97-112.
33. Broecker, W.S., and Van Donck, J. : 1970, Rev. Geophys. Space Phys. 8, pp. 169-198.
34. Cathles, L.M. : 1975, "The Viscosity of the Earth's Mantle", Princeton Univ. Press., New Jersey, 386 pp.
35. Peltier, W.R. : 1981, Ann. Rev. Earth Planet. Sci. 9, pp. 199-225.
36. Burgers, J.M., and Collette, B.J. : 1958, Proc. Kon. Ned. Akad. Wet. B, 61, pp. 221-241.
37. Thomas, R.H., and Bentley, C.R. : 1978, Quaternary Research 10, pp. 150-170.
38. Thomas, R.H. : 1977, Geogr. Phys. Quat. 31, pp. 347-356.
39. Heath, G.R. : 1979, Palaeogeog., -climatol., -ecol. 26, pp. 291-303.
40. Chappell, J., and Veeh, H.H. : 1978, Bull. Geol. Soc. Amer. 89, pp. 356-368.
41. Berger, A.L. : 1978, Quaternary Research 9, pp. 139-167.
42. Shackleton, N.J., and Opdyke, N.D. : 1973, Quaternary Research 3, pp. 39-55.
43. Emiliani, C. : 1978, Earth Planet. Sci. Lett. 37, pp. 349-352.
44. Imbrie, J., Hays, J.D., Martinson, D.G., McIntyre, A., Mix, A.C., Morley, J.J., Pisias, N.G., Prell, W.L., and Shackleton, N.J. : 1984, in : "Milankovitch and Climate", A. Berger, J. Imbrie, J. Hays, G. Kukla, B. Saltzman (Eds), Reidel Publ. Company, Holland. This volume, p. 269.

[1] Present address : 3067 Bateman Street, Berkeley, Cal. 94705, USA.

[2] It is assumed here that all deglaciations of the last 10^6 years were as complete as that of today, as was probably the case during the last interglacial at 125 kyr BP However, there is some uncertainty from available data as to exactly how far the ice sheets retreated during earlier interglacials (12, 40).

A MODEL OF THE ICE AGE CYCLE

W.R. PELTIER and W. HYDE

Department of Physics, University of Toronto,
Toronto, Ontario, M5S 1A7 Canada.

ABSTRACT

Time series analyses of the $\delta^{18}O$ signal in deep sea sedimentary cores have very clearly established that ice volume fluctuations throughout the late Pleistocene have been dominated by an almost periodic oscillation with a characteristic timescale of 10^5 years. Although statistically significant variance is also found at the expected Milankovitch periods corresponding to the precession of the equinoxes and to the variation of orbital obliquity, the dominance of the 10^5 year signal has been difficult to understand because the strength of the astronomical forcing at this period is negligibly small. The analysis presented here shows that this period arises naturally in a new model which includes the nonlinearity due to ice sheet flow and a more accurate description of the process of glacial isostatic adjustment than has been employed previously.

INTRODUCTION

One of the most intriguing problem in the geophysical sciences which has re-emerged in recent years concerns the origin of ice age and the attempt to understand the detailed characteristics of the ice age in which we are currently living. For some time this problem was addressed rather exlusively by members of the atmospheric sciences community whose investigations focussed upon efforts to determine the magnitude of the decrease in solar insolation which would be required to initiate the transition from an essentially ice free climate similar to today's, to one in which a large fraction of the continental surface was ice covered. These investigations were originally based on the use of energy balance climate models of the Budyko-Sellers type (e.g., (1)) and the results obtained with such models have been recently reviewed in North, Cahalan and Coakley (2). Both these simple energy balance climate models and the more all inclusive general circulation models (e.g., (3)) suggest that a decrease of the solar constant of 5-10% from present day values would be required to initiate an ice covered earth.

Research in this area was given a new injection of vitality by the paper of Hays, Imbrie and Shackleton (7) in which $\delta^{18}O$ data from deep sea sedimentary cores were employed to demonstrate rather conclusively that the climate system was responding to the small changes of solar insolation received by the Earth due to temporal variations in the geometry of its orbit around the sun. These temporal variations are known to occur at characteristic period of 19 000 and 23 000 years due to the precession of the equinoxes, at 41 000 years due to the variation of orbital obliquity, and at 100 000 and 400 000 years due to the change of orbital eccentricity. Although the latter source of variability is associated with a maximum change of only 0.2% in the total received radiation, the changes in orbital eccentricity strongly modulate the amplitude of the precession effect such that insolation anomalies at times of high eccentricity are 3-4 times greater than at times of low orbital eccentricity. For 60° north latitude this effects a variation of summer seasonal insolation which is on the order of 8% about the mean value with the perturbation at summer solstice exceeding 13% of the mean (9,10). In the original astronomical theory of the ice ages, Milankovitch (11) suggested also that the variation of summer insolation in these latitudes might be the crucial determinant of the inception of northern hemisphere continental glaciation, although he used only the insolation of the caloric seasons. The contribution of the paper by Hays et al. (7) was to establish that these small insolation fluctuations were inducing large responses of the planetary cryosphere. A useful recent review is that by Crowley (12).

Several examples of the type of information analyzed by Hays et al. are reproduced in Figure 1 which is based upon data from Imbrie et al. (4) and Shackleton and Opdyke (5, 6) as originally composited in Oerlemans (13). This shows the variations of the isotopic ratio $^{18}O/^{16}O$ as a function of depth in several typical long sedimentary cores. Following analysis in Shackleton (29), Imbrie and Kipp (30), Shackelton and Opdyke (5), it is generally accepted that this ratio is direct proxy for continental ice volume. The depths marked M-B on three of the four examples shown denotes the depth of occurence of the last reversal of the Earth's magnetic field which may be located

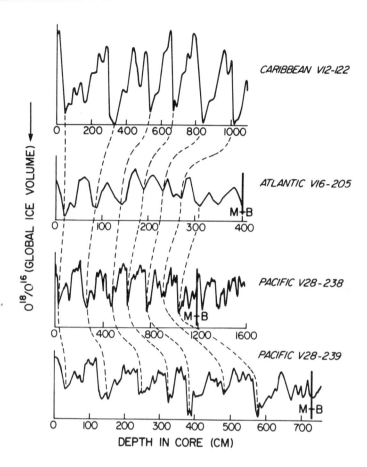

Figure 1 A comparison of oxygen isotope records from four different deep sea sedimentary cores based upon data from Imbrie et al. (4) and Shackleton and Opdyke (5,6). The heavy vertical bars marked M-B denote the depth corresponding to the Matuyama-Brunhes boundary of age 730 (±20) kyr.

in the core by scanning with a magnetometer. Based upon isotopic age determinations of this feature in volcanic sequences on land, it is well known that the age of this horizon is 730 000 ± 20 000 years (14). Using the depth-time connection provided by the location of the horizon and the assumption of a constant sedimentation rate, the $\delta^{18}O$ vs. depth data may be converted into a time series to which standard analysis techniques may be applied. Power spectra for the last two cores shown on Figure 1 are shown on Figure 2, which is reproduced from (8). These spectra very clearly reveal the

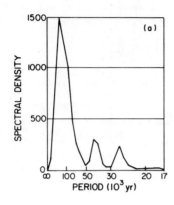 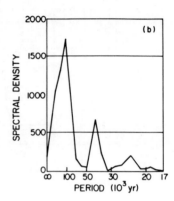

Figure 2 Power spectra of the $\delta^{18}O$ time series from Paific cores V28-238(a) and V28-239(b) reproduced from Birchfield et al. (8). Note the dominance of the spectral peak at 10^5 years period.

features first discovered by Hays et al. (7) which consist of statistically significant spectral peaks corresponding to periods near 23 000, 41 000 and 100 000 years. These data very clearly establish the sensitivity of the cryosphere to the orbital forcing.

They do not, however, support the notion that the response to the orbital forcing at each of the astronomical periods is simply proportional to the summertime insolation anomaly at that period. Figure 3 shows a power spectrum of the insolation anomaly at 60°N latitude computed from the paleo-insolation time series of Berger (15). Although there is substantial power at the precession and obliquity periods, that at 10^5 years is completely negligible in comparison. Since the dominant oscillation of the cryosphere is at this period, as evidenced by Figure 2, it is clearly impossible to understand the reason for the existence of the "ice age cycle" in terms of a linear representation of the climate system. No fully satisfactory explanation of this 10^5 year oscillation, which was first discovered by Broecker and Van Donk (16), has yet been proposed (17). The purpose of this paper is to illustrate a new nonlinear model which we have constructed to address this question. The model appears to deliver the 10^5 year cycle in a completely straightforward way.

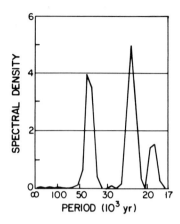

Figure 3 Power spectrum of the insolation time series of Berger (15) for 60° north latitude summer.

A SPECTRAL MODEL OF THE 10^5 YEAR PLEISTOCENE CLIMATE OSCILLATION

Although there have been several different models recently proposed to resolve the problem posed by the above described data, including that by Ruddiman and McIntyre (18) which relies on the phase relationship between precession and obiquity forcing, and that by Saltzman et al. (19) in which the deep ocean circulation plays a prominent role, these models are at best semiquantitative and therefore not fully satisfactory. More directly relevant, in our opinion, are the models originated by Weertman (20) in which the nonlinearity associated with the flow of ice sheets is explicitly included. Of particular interest to us will be models in which the physics of ice sheet flow are coupled through an insolation forced accumulation function to the physics of the sinking of the earth under the weight of the ice. Although very simple, so-called zero dimensional, climate models have been designed wich contain this physics (21, 22) and appear to be able to explain the appearance of an oscillation with a period of 10^5 years, they do so by the nonlinear generation of combination tones of the 19 kyr and 23 kyr periods of the precessional forcing as first suggested by Wigley (23). It seems clear that the strength of the combination tones is liable to be an artifact of the fact that these models contain no explicit spatial dependence and therefore that this explanation of the 10^5 year cycle is rather suspect. At the very least it is an explanation which must be tested in the context of models which explicitly resolve the spatial structure of the ice sheets.

The first models of which we are aware that contained explicit representations of the physical processes mentioned above, namely ice sheet flow and accumulation and glacial isostatic adjustment, were those published by Oerlemans (13, 24) and Birchield et al. (8). The analysis presented in these two papers was based upon essentially identical mathematical models, yet the conclusions reached on the basis of the respective model calculations were diametrically opposite. In Peltier (25) it was first pointed out that the only fundamental difference in the two models was the timescale assumed for the process of glacial isostatic adjustment. Oerlemans assumed that this timescale was unknown and so varied it to give the best simulation of the 10^5 year oscillation, an approach which let to a required adjustment timescale τ of between 10 000 and 30 000 years. His model simulations of the insolation forced ice volume fluctuations were then dominated by an oscillation with period near 10^5 years but the result was quite sensitive to his choice of the model parameters and consequently was unsatisfactory as a convincing demonstration that this combination of physical processes provided the explanation of the observed cycle. The published analysis by Birchfield et al. (8) employed a fixed adjustment timescale of 3 000 years and showed that with this choice of parameter the power in the cryosphere oscillation at 10^5 period was extremely small, in accord with the findings of Oerlemans. This analysis severely undermined Oerlemans conclusion that the 10^5 year cycle could be due to the interplay between ice sheet flow and glacial isostatic adjustment, since Birchfield et al. correctly argued that the adjustment timescale required by relative sea level data from within the margin of the Laurentian ice sheet was only 2 kyr whereas that from Fennoscandian was 4 kyr. Their choice of 3 kyr was therefore a compromise between the requirements of the data from these two main centres of glaciation and was too short to allow the oscillatory behaviour obtained by Oerlemans to occur. Peltier (25) suggested that the difficulty might be resolved if a more realistic representation of the glacial isostatic adjustment process were implemented in the model and our purpose here will be to demonstrate that this modification of the model allows for an extremely straightforward reproduction of the 10^5 year cycle.

The model which we will employ is the spectral model whose mathematical structure was derived in Peltier (25). The use of a spectral model in this application is dictated by the nature of the physical process of glacial isostatic adjustment as we will see. Although the basic model may be expanded to include the description of ice sheet growth and decay on a spherical surface with a realistic distribution of oceans and continents, we will here restrict consideration to the simple axially symmetric geometry illustrated in Figure 4 in which the ice sheet consists of a ring of ice surrounding a single polar ocean. The

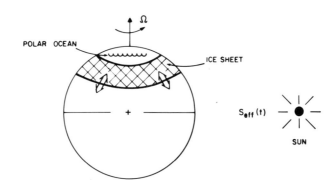

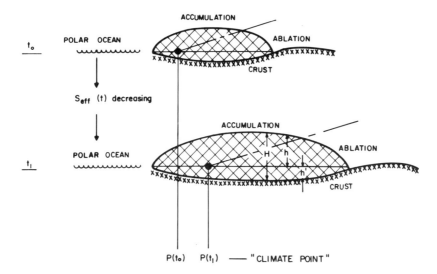

Figure 4 Schematic diagram for the paleoclimatic model which consists of an active ice sheet driven by variations of effective solar insolation and modified by the influence of the sinking of the earth under the weight of the ice.

expansions and contractions of the ice sheet are described by the vertically integrated continuity equation for ice which takes the form :

$$\frac{\partial H}{\partial t} = -\frac{1}{r \sin\theta} \frac{\partial}{\partial \theta} (\sin\theta U) + A(\theta, t), \qquad [1]$$

in which the ice flux U is determined by the Glen flow law (e.g. 26) as :

$$U = \frac{1}{5} \hat{c} \, (\rho g)^3 H^5 \left[-\frac{1}{r} \frac{\partial h}{\partial \theta} \right]^3 . \qquad [2]$$

In [1], H = h + h' is the ice sheet thickness in which h is the height above and h' the depth below sea level, $\hat{c}$ is an empirical constant (e.g. 8), ρ is ice density, g the surface gravitational acceleration, and A(θ,t) is the accumulation function through which the variations of solar insolation effect fluctuations of ice volume. Equation [1] may be cast into the form of a nonlinear diffusion equation as :

$$\frac{\partial h}{\partial t} = \frac{1}{r \sin\theta} \frac{\partial}{\partial \theta} \left[\sin\theta K \frac{\partial h}{\partial \theta} \right] - \frac{\partial h'}{\partial t} + A(\theta,t) \qquad [3]$$

in which the nonlinear diffusion coefficient which determines the rate of flow of the ice sheet is

$$K = \frac{1}{5} \hat{c} \, (\rho g)^3 \frac{H^5}{r^3} \left[\frac{\partial h}{\partial \theta} \right]^2 . \qquad [4]$$

To complete the model we need a separate equation for the dependent variable h'. In the modern theory of glacial isostasy reviewed in (25), h' is given by the convolution of the surface load H with a viscoelastic Green's function u_r as :

$$h'(\theta,\lambda,t) = \int dt' \int \int d\Omega' u_r(\theta/\theta',\lambda/\lambda',t/t') \rho_I H(\theta',\lambda',t') \qquad [5]$$

The Green's function $u_r(\theta,\lambda,t)$ is obtained by solving the appropriate boundary value problem for a radially stratified viscoelastic model of the planetary interior. This analysis gives u_r as the sum of elastic and viscous parts of the form

$$u_r(\theta,t) = u_r^E(\theta)\delta(t) + \frac{a}{m_e} \sum_{l=0}^{\infty} \sum_{j=1}^{M} r_j^l e^{-s_j^l t} P_l(\cos\theta) \qquad [6]$$

where u_r^E is the elastic (instantaneous) part and the double summation gives the viscous contribution. In the latter r_j^l and s_j^l are the initial amplitudes and inverse decay times of the M normal modes of viscous gravitational relaxation which are required to synthesize the temporal behaviour of the degree l component of the deformation spectrum. When [6] is substituted in [5] and the equation differentiated with respect to time we obtain the following evolution equation for the h_l':

$$\frac{\partial h_l'}{\partial t} = \frac{a}{m_e} \cdot \frac{4\pi}{(2l+1)} \cdot \sum_j (-r_j' s_j') e^{-s_j' t} \int_{-\infty}^{t} dt' \, e^{s_j' t'} H_l(t') a^2 \rho_I$$

$$+ \frac{a}{m_e} \cdot \frac{4\pi}{(2l+1)} \cdot \sum_j r_j' H_l(t) a^2 \rho_I + \frac{a}{m_e} \cdot \frac{4\pi}{(2l+1)} \cdot q_l^E \frac{dH_l}{dt} a^2 \rho_I, \quad [7]$$

in which q_l^E has been used to denote the elastic Love number of degree l (25). If we make the approximations : (i), that the elastic component of the response to loading may be neglected and (ii), that the temporal behaviour of each harmonic may be described by a single exponential relaxation with the inverse decay time s_l', then [7] reduces to the much simpler form

$$\frac{\partial h_l'}{\partial t} = C_l r^l h_l + (-s^l + C_l r^l) h_l' \quad [8]$$

where

$$C_l = \frac{a}{m_e} \cdot \frac{4\pi}{(2l+1)} \cdot \rho_I a^2 .$$

in which a, m_e and ρ_I are the earth's radius and mass and ρ_I is the density of ice. The general form of [8] which includes both elastic and realistic multimode effects is discussed in detail in (25).

In order to complete the formulation of the model we require a spectral form for the equation of ice sheet flow [3]. To obtain this we expand $H = h + h'$, K, and A in terms of Legendre polynomials and substitute in [3]. If the equation is then multiplied by P_l and integrated from $\cos \theta = -1$ to $\cos \theta = +1$ we obtain the following spectral form of the nonlinear diffusion equation as :

$$\frac{\partial h_l}{\partial t} = B_{lmn} K_m(t) h_n(t) - \frac{\partial h_l'}{\partial t} + A_l(t) , \quad [9]$$

in which the interaction matrix B_{lmn} is

$$B_{lmn} = \frac{-(2l+1)}{2a} \int_{-1}^{+1} dx\, P_l(x) \left[\frac{x}{(1-x^2)^{1/2}} P_m(x) P_n(x) \right.$$

$$\left. + P_m'(x) P_n'(x) + P_m(x) P_n''(x) \right], \qquad [10]$$

which has elements which depend only on the basis functions P_l and which may therefore be computed once and for all. In the remainder of this paper we will discuss solutions for the approximate spectral model which is described by eqs [8] and [9] and which has dependent variables h_l and h_l'. In order to run the model we must specify the accumulation function A which we shall assume is representable as in Birchfield et al. (8) in the form

$$A = a(1-bh)\, 0, \text{ above the firn line} \qquad [11]$$

$$= a'(1-bh)\, 0, \text{ below the firn line}$$

where the firn line separates the region of accumulation from the region of ablation and is illustrated on Figure 4. Forcing due to the anomaly in solar insolation is introduced into the model by moving the climate point (Fig. 4) southward during times of negative anomaly and northward during times of positive anomaly. As in (8) we will compute the shift in the latitude of the climate point from:

$$\partial x = -C \partial Q \qquad [12]$$

where ∂Q is the insolation anomally and $C=1/(dQ/dx)$ is fixed by the present insolation gradient. For all of the calculations we will describe the parameters have been fixed close to the values selected by (8) as:

$$\begin{aligned} C &= 43.35 \text{ W/m}^2 \text{ km} \\ a &= 1.2 \text{ m/yr} \\ a' &= -3.6 \text{ m/yr} \\ b &= 2.3 \times 10^{-4}/\text{m} \end{aligned} \qquad [13]$$

Although calculations which will be described elsewhere use the insolation anomaly time series of Berger (15) to force the model system, we will focus here on calculations designed to isolate the features of the model which are crucial to the oscillation and so will investigate solutions in which the forcing is through a synthetic insolation time series which consists of a pure sinusoid with period 20 000 years which is meant to represent the precessional forcing which is still dominant even at 60°N latitude.

RESULTS OF AN INITIAL MODEL EXPERIMENT

The model described above differs physically from those of Oerlemans (24) and Birchfield et al. (8) in that it contains a much more accurate description of the process of glacial isostatic adjustment. The main characteristics of the realistic models of glacial isostasy which have been developed over the past decade of work which began with Peltier (27) and which has been most recently reviewed in (25) is that each wavenumber in the spectrum of the surface deformation is characterized by a unique set of exponential relaxation time s_j^l. Figure 5 shows the relaxation spectrum for a typical earth model

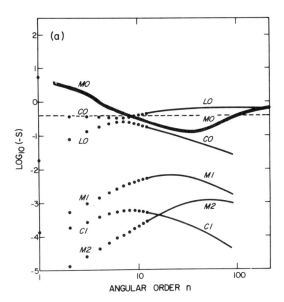

Figure 5 Relaxation spectrum for a typical viscoelastic earth model. Neglecting the inner core mode C1 which is spurious such models have effectively 5 important modes of viscoelastic relaxation. The dominant mode M0 is supported by the density discontinuity across the free outer surface of the model and is set off on the Figure by the heavy solid line. The dashed line denotes the flat spectrum (relaxation time independent of angular order) employed in the analyses of Oerlemans (24) and Birchfield et al. (8).

which fits a large fraction of the observables of glacial isostasy including relative sea level, free air gravity, and earth rotation observations. The model has a constant mantle viscosity of 10^{21} Pas and a lithospheric thickness of 120 km. Inspection shows that realistic earth models have essentially five different modes of relaxation for each spherical harmonic degree, each of which is supported by a specific physical characteristic of the radial viscoelastic structure of the planet. The dominant mode M0 is supported by the density jump across the free outer surface of the planet, C0 by that across the core mantle boundary, L0 by the infinite viscosity contrast across the lithosphere-mantle interface, and the pair M2, M1 by the density jumps across the Olivine-Spinel and Spinel-post Spinel phase transitions at 420 and 670 km depth in the Earth. The latter modes exist only if the phase reaction kinetics are slow, however, or the boundaries are in fact chemical in origin. Since we shall employ the approximate model embodied in equations [8] and [9] for the calculations to be described here we are obliged to truncate the spectrum shown on Figure 5 to a single mode for each spatial harmonic. Since the M0 mode is strongly dominant for all $l \geqslant 10$ (28) we shall restrict attention to this modal branch and fix the r^l, corresponding to these s^l, by the requirement that all of the viscous relaxation be carried by this single mode.

On Figure 5 we have set the M0 branch off from the others by drawing a solid line through the modes which comprise it. The dashed line on this Figure shows the relaxation spectrum employed by Birchfield <u>et al.</u> (8) for comparison. Their spectrum consists of a constant relaxation time of 3 000 years for all waves numbers. Inspection of Figure 5 shows that the variation of relaxation time with l for a realistic model is strong, however, and such that relaxation times for a wide range of l values are much longer than the 2000 years visible in the sea level record from sites near the centres of rebound in Hudson Bay and the Gulf of Bothnia. We will show here that the "stiffness" of the wavenumber spectrum for a realistic earth model, even when the long timescale extra modes are neglected, is the physical feature of the model which allows it to deliver not only the correct 10^5 year period of the ice volume oscillation inferred from $\delta^{18}O$ data but also explains the very rapid terminations of each glacial cycle. In (25) it was suggested that the very long timescale modes M1 and M2 may have been required to provide the necessary stiffness. Explicit calculations have now established that this degree of complexity is in fact unnecessary.

Figure 6a shows the synthetic ice volume time series obtained by solving the model system [8], [9] and [11] with 20 000

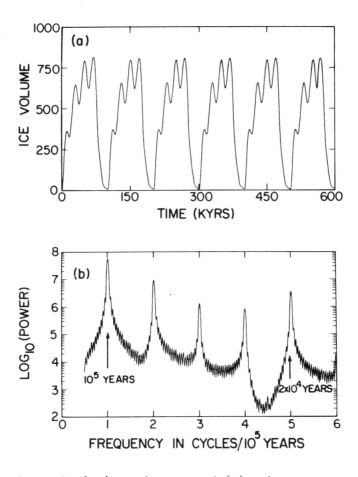

Figure 6 Synthetic ice volume record (a) and power spectrum of the synthetic (b) from the paleoclimate model consisting of equations [8] and [9]. The ice volume record is clearly dominated by a periodic oscillation of period 10^5 years. The individual 10^5 year pulses are characterized by a slow oscillatory rise to maximum volume and an extremely rapid collapse to zero volume. These fast "terminations", which are charactaric of the 10^5 year cycle, have not previously been explained without recourse to rather ad hoc alterations of the model-to mimic for example the effect of "calving" into proglacial lakes. The nature of these terminations is further discussed in the paper by Wallace Broecker in this volume. Volume units in (a) are $m^3/10^{14}$.

year sinusoidal insolation anomaly forcing. The response to this forcing is clearly dominated by a periodic signal with a period very close to 10^5 years and a pulse shape characterized by a relatively slow glaciation phase followed by an extremely rapid termination lasting on the order of 10-15 kyr. Riding on the dominant 10^5 year oscillation is a 20 000 year periodic signal which represents the direct response of the model at the forcing period. Cross spectral analysis shows that this component of the response lags the forcing by approximately 5 000 years in accord with the phase lag measured from the $\delta^{18}O$ signal by Hays et al. (7). Figure 6b shows the power spectrum of a synthetic ice volume time series consisting of twenty 10^5 year cycles and this line clearly dominates that at the forcing period as is the case with the observed spectra, examples of which were shown on Figure 2. Also clearly evident on Figure 6b are spectral lines at each of the harmonics of the basic 10^5 year period. Such lines are not observed in the real geological data but this is certainly due to the fact that the pulse shape of the actual cryospheric volume oscillation is not exactly preserved from cycle to cycle (see Fig. 1). Aside from these relatively minor details we feel justified in claiming that our spectral model of the Pleistocene climatic oscillation is very successful in replicating all of the important features of this physical phenomenon.

CONCLUSIONS

In spite of the extraordinary success of this simple model in mimicing the observed signal there are obviously several further numerical experiments which must be performed with it before we will be in a position to claim that our physical model is the correct model of these observations. Foremost among these are realistic Milankovitch experiments which employ the complete insolation time series of Berger (9) to force the ice sheet. We are also obliged to demonstrate the stability of our nonlinear oscillation against stochastic perturbations in order that we may be confident that the physical process we have isolated is insensitive to the presence of realistic levels of random noise. We also intend to investigate the question of whether the oscillation is predictable in a model with a realistic distribution of continents and oceans by coupling it to a simple energy balance model of atmospheric climate which resolves these surface features. On basis of the results reported here we are rather optimistic that this is feasible program and that the physical processes which we have isolated do provide the correct explanation for the 100 000 year oscillation of the Pleistocene cryosphere.

REFERENCES

1. Sellers, W.D. : 1965, Physical Climatology. Univ. of Chicago, Illinois.
2. North, G.R., Cahalan, R.F. and Coakley, J.A. : 1981, Rev. Geophys. Space Phys., 19, pp. 91-121.
3. Wetherald, R.T. and Manabe, S. : 1975, J. Atmos. Sci., 3, pp. 2044-2059.
4. Imbrie, J., Van Donk, J. and Kipp, N. : 1973, Quat. Res., 3, pp. 10-38.
5. Shackleton, N.J. and Opdyke, N.D. : 1973, Quat. Res., 3, pp. 39-55.
6. Shackleton, N.J. and Opdyke, N.D. : 1977, Nature, 270, pp. 216-219.
7. Hays, J.D., Imbrie, J. and Shackleton, N.J. : 1976, Science, 194, pp. 1121-1132.
8. Birchfield, G.E., Weertman, J. and Lunde, A.T. : 1981, Quat. Res., 15, pp. 126-142.
9. Berger, A.L. : 1978, J. Atmos Sci., 35, pp. 2362-2367
10. Berger, A.L. : 1979, Il Nuovo Cimento 2C(1), pp. 63-87.
11. Milankovitch, M. : 1941, Canon of insolation and the ice age problem. K. Serb. Acad. Beorg. Spec. Publ., 132.
12. Crowley, T.J. : 1983, Rev. Geophys. Space Phys., 21, pp. 828-877.
13. Oerlemans, J. : 1980, Nature, 287, pp. 430-432.
14. Cox, A., and Dalrymple, G.B. : 1967, J. Geophys. Res., 72, pp. 2603-2614.
15. Berger, A.L. : 1982, Quat. Res., 9, pp. 139-167.
16. Broecker, W.S. and Van Donck, J. : 1970, Rev. Geophys. Space Phys., 8, pp. 169-198.
17. Imbrie, J. and Imbrie, J.Z. : 1980, Science, 207, pp.943-953.
18. Ruddiman, W.F. and McIntyre, A. : 1981, Quat. Res., 16, pp. 125-134.
19. Saltzman, B., Sutera, A. and Evenson, A. : 1981, J. Atmos. Sci., 38, pp. 494-503.
20. Weertman, J. : 1976, Nature, 261, pp. 17-20.
21. Kallen, E., Crafoord, C. and Ghil, M. : 1979, J. Atmos. Sci., 36, pp. 2292-2303.
22. Ghil, M. : 1981, in Climate Variations and Variability : Facts and Theories, edited by A. Berger, pp. 539-557. D. Reidel, Hingham, Mass.
23. Wigley, T.M.L. : 1976, Nature, 264, pp. 629-631.
24. Oerlemans, J. : 1981, Tellus, 33, pp. 1-11.
25. Peltier, W.R. : 1982, Adv. in Geophys., 24, pp. 1-146.
26. Paterson, W.S.B. : 1981, The Physics of Glaciers. Pergamon, Oxford.
27. Peltier, W.R. : 1974, Rev. Geophys. Space Phys., 12, pp. 649-669.
28. Wu, Patrick and Peltier, W.R. : 1982, Geophys. J. Roy. Astr. Soc., 70, pp. 435-486.

29. Shackleton, N.J. : 1967, Nature, 215, pp. 15-17.
30. Imbrie, J. and Kipp, N.J. : 1971, in The Late Cenozoic Glacial Ages, edited by K.K. Turekian, pp. 71-179. Yale University Press, New Haven, Conn.

SENSITIVITIES OF CRYOSPHERIC MODELS TO INSOLATION AND TEMPERATURE VARIATIONS USING A SURFACE ENERGY BALANCE

T.S. Ledley[1]

Department of Meteorology and Physical Oceanography, Massachusetts Institute of Technology, Cambridge, Massachusetts, USA

ABSTRACT

A zonally averaged dynamic continental ice sheet model and a thermodynamic sea ice model are used to examine the sensitivity of the cryosphere to variations in insolation and temperature on Milankovitch time scales. These models use an energy balance parameterization of the net budget of accumulation and ablation. Continental ice sheet model results indicate that, with the present zonally averaged energy balance calculation, an ice sheet, once formed, fluctuates little in latitudinal extent. Sea ice model results indicate the magnitude of the energy exchange between the ocean and atmosphere can vary greatly between ice-covered and ice-free conditions.

INTRODUCTION

Models of ice sheet and sea ice are important for understanding the mechanisms of long-term glacial cycles. Previous studies, emphasizing ice sheet and coupled ice sheet-climate models, have produced 20 kyr, 40 kyr, and 100 kyr climatic variations similar to those seen in the geological record. The simple parameterizations of accumulation and ablation used in these studies do not explicitly describe specific physical processes. Consequently, it is not clear, from their model results, what the important physical processes are in large-scale glacial variations.

In this study models of both sea ice and continental ice, incorporating an energy balance parameterization of accumulation and ablation (the net budget), are used to examine cryospheric responses to solar insolation and temperature fluctuations on Milankovitch time scales.

This report describes preliminary experiments with the cryospheric models using both energy balance calculations and earlier parameterizations of net budget. Sensitivity studies are performed varying climatic parameters and insolation in order to study the response of these models in the hope of finding more quantitative, physical explanations for observed glacial volume changes.

MODEL FORMULATIONS

The Energy Balance Net Budget

The continental ice sheet and sea ice models are forced through energy and mass exchanges at their interfaces with the atmosphere. The energy exchanges include short-wave radiation absorbed at the surface, incoming long-wave radiation from the atmosphere, outgoing long-wave radiation from the surface, sensible heat, latent heat, and conduction of heat from below the surface. In the continental ice sheet model the energy balance is modified by neglecting terms for the conductive flux from below the surface of the ice sheet and the penetration of short-wave radiation into the ice sheet.

The mass exchanges at the ice sheet and sea ice surfaces take into account the accumulation and ablation of ice and snow. Accumulation is parameterized as a function of present precipitation, air temperature, and saturation vapor pressure. Ablation is computed using the energy balance condition. The surface snow/ice temperature is allowed to vary freely in response to the energy balance as long as it remains below the freezing point, 271.2 K. Once the surface temperature reaches the freezing point any additional energy available for warming is used to ablate the ice surface.

Incoming Solar Radiation

The determination of incoming solar radiation is done in two steps. The radiation at the top of the atmosphere as a function of latitude and time is computed following Berger (1). Then the fraction of that radiation reaching the surface after being modified by the atmosphere is determined.

The short-wave radiation reaching the surface is computed using an adaptation of the planetary albedo parameterization developed by Thompson (2) and Thompson and Barron (3). The solar radiation at the top of the atmosphere is attenuated through absorption by ozone, scattering and absorption by water vapor and dry aerosols, and Rayleigh scattering. The direct and diffuse radiation under clear skies is then determined from the remaining radiation and the direct and diffuse surface albedos. The albedos are set empirically depending on surface type - sea ice, open ocean, land, or continental ice - and are modified by zenith angle. Under cloudy skies the short-wave radiation reaching the surface is further modified by the cloud albedo, absorption by the cloud, and absorption and multiple reflections between the cloud base and surface. The total solar radiation available at the surface is a linear combination of that under clear skies and cloudy skies weighted by the cloud fraction.

The Sea Ice Model

The sea ice model is a modification of that used by Semtner (4). It is a three-layer thermodynamic model which includes conduction within the ice and snow, penetration of solar radiation into ice layers, and surface energy balances. This model is coupled to a very simple mixed-layer ocean model which allows the initiation of new ice in open ocean, and the inclusion of a lead parameterization.

Sea ice is assumed to be a uniform horizontal slab on which snow may accumulate. The temperatures within snow and ice are governed by similar one-dimensional heat equations. The surface temperature is governed by the energy balance condition.

In addition to the energy balance applied at the atmosphere-ice interface an energy balance is applied at the ice-ocean interface. This balance involves the conductive flux within the ice and the upward flux from the ocean. Any imbalance in these fluxes is offset by the accretion or ablation of ice.

The transition from ice-free to ice-covered ocean is made by assuming an isothermal oceanic mixed layer exists below the ice. If ice disappears a heat budget is applied to this layer until the ocean again reaches the freezing point, at which time new ice forms.

Sea ice has a dramatic insulating effect on the transfer of energy between the ocean and atmosphere (5). This effect is modified by the presence of leads, or cracks, in the ice pack. Leads are generally caused by stresses put on the ice slab by winds and ocean currents. These dynamic mechanisms are not included in this model but are accounted for by a simple lead parameterization.

The Continental Ice Sheet Model

The continental ice sheet model is made up of a zonally averaged dynamic model of ice flow with specified topography, time-lagged bedrock depression, and a surface energy balance that determines the net budget of accumulation and ablation at the surface of the ice.

The ice flow model determines the horizontal continental ice velocity and ice thickness as a function of latitude and time. It was derived from the equations of motion, the Generalized Flow Law (6), and Glen's Creep Law (7). The model is similar to that used by Pollard (8) and Birchfield et al. (9). The model equations compute the vertically averaged velocity, the time rate of change of the land height, and the time rate of change of the ice thickness. Other variables and parameters which are included in the model equations are the flow parameters, A and n, where A is generally set to 0.15 bar^{-3} yr^{-1} and n is set to 3, the equilibrium land height, the ice-free equilibrium land height, the reciprocal of the e-folding time of the bedrock response, which is generally set to $1/3000$ yr^{-1} in these experiments, and the net budget of accumulation and ablation. The model equations are applied to a one-degree latitudinal grid.

The source of a continental ice sheet is the accumulation of snow at its surface. In order to determine the net budget, the accumulation minus the ablation which occurs over the time step of integration of the dynamic ice flow equations, it is necessary to compute the total amount of snowfall and ablation over the time step. This is done using the surface energy balance equation described earlier. Here the net budget, computed at two-week seasonal time steps, is summed over the 10-year ice sheet time step to produce a net budget. Energy balance calculations are carried out on a five-degree latitudinal grid and interpolated to the finer one-degree latitudinal ice sheet grid.

In addition to the ablation computed from the surface energy balance equation, there is ablation due to calving of the ice into the oceans at the edges of the continents. As this ice is no longer part of the dynamic flow of ice on the continent it is treated as an ablation and added to the net budget. Regions of calving ablation occur at the southern edge of the Arctic Ocean and at the northern edge of Antarctica.

Once the complete net budget is determined for the ten-year ice sheet time step it is used to determine the new ice sheet thickness.

Further details concerning the formulation of the models and the energy balance net budget calculations can be found in Ledley (10).

ICE SHEET MODEL SENSITIVITY EXPERIMENTS

Equilibrium Line Net Budget

Experiments were done to test the correspondence between the response of the ice sheet model used here and that employed by Pollard (8). This was done using Pollard's net budget parameterization which involves defining an equilibrium snow-fall line which changes elevation in response to variation in the summer half-year insolation at 55°N. The results of this test, with the flow parameter, A, set to 2.1 bar^{-3} yr^{-1} (the value used by Pollard), is shown in Figure 1. This corresponds well with the results shown in Pollard's Figure 1 (8). The only real difference is that the cross-sectional area of this ice sheet is smaller by as much as a factor of two. This result indicates that the response of the ice model used here is comparable to that used by Pollard.

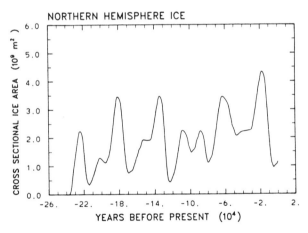

Figure 1 Northern hemisphere cross-sectional ice area from -240 kyr to the present using the equilibrium line net budget parameterization, A equal to 2.1 bar^{-3} yr^{-1}, and the bedrock depression constant equal to 1/3000 yr^{-1}.

The above experiment was repeated with the flow parameter, A, set equal to 0.15 bar^{-3} yr^{-1}, which is in the range more commonly used in ice sheet models and is in general agreement with measurements in Antarctica and Greenland. This change decreased the ice flow to the extent that snowfall at the surface built up the ice sheet to unrealistic elevations, and eventually exceeded the stability limits of the model.

Energy Balance Net Budget Experiments

In this study the following simple parameterization linking air temperature and, therefore, the net budget to Milankovitch insolation variations is used.

$$T_a = (Q(1-\alpha_p) - Q_b(1-\alpha_{pb}))/B$$

Q is the insolation at the top of the atmosphere, α_p is the planetary albedo, T_a is the surface air temperature, B is the sensitivity parameter of air temperature to insolation variations, and the subscript "b" indicates the present. This equation states that the change in surface air temperature is proportional, through the sensitivity parameter, to the change in the absorbed solar radiation from the present. The value of B is uncertain. Results from a study by Warren and Schneider (11) indicate that for a global equilibrium the value of B is about 2 $W m^{-2} K^{-1}$. Since the time scale of the energy balance calculation is seasonal and is done at individual latitudes a larger value of B for the seasonal application would seem more appropriate.

Experiments were performed in which present conditions (no northern hemisphere ice sheets) were used as initial conditions and the model was run with Milankovitch-varying insolation forcing the energy balance net budget. It was found that an ice sheet would not grow under those conditions even if the mean air temperature was dropped by ten degrees. However, an ice sheet would begin to grow if the experiment was initialized with a thin snow field with an albedo of 0.8 (an ice surface albedo of 0.6 was used for all other experiments in this report) and the mean air temperature was dropped only three degrees. The introduction of year-round high albedo caused a sufficient reduction in the amount of short-wave energy absorbed at the surface to permit the snow field to be maintained during the summer season, and as a result allow the ice sheet to grow further during the winter. This result is supported by the work of North et al. (12) found elsewhere in this volume.

In order to test the sensitivity of a zonally averaged ice sheet to variations in insolation and air temperature once the ice sheet is formed the model was initialized with the ice sheet profile at -48 kyr in the equilibrium line experiment and integrated for 58 kyr with values of B ranging from 4 to 12 $W m^{-2} K^{-1}$, and the flow parameter, A, set to 0.15 $bar^{-3} yr^{-1}$.

The results, shown in Figure 2, indicate that the ice sheet is very sensitive to the value of B. When B is set to 4 W m^{-2}K^{-1}

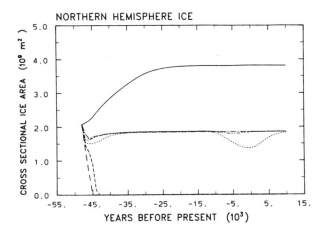

Figure 2 Northern hemisphere cross-sectional ice area from -48 kyr to +10 kyr computed from the energy balance net budget parameterization for A equal to 0.15 bar^{-3} yr^{-1} and B equal to 4 W m^{-2}K^{-1} (solid line), 6 W m^{-2}K^{-1} (single-dot dashed line), 8 W m^{-2}K^{-1} (double-dot dashed line), 9 W m^{-2}K^{-1} (dotted line), 10 W m^{-2}K^{-1} (short dashed line), and 12 W m^{-2}K^{-1} (long dashed line).

the ice sheet grows from its initial size, and comes to an approximate equilibrium which seems to be independent of insolation and temperature variations. The growth over the first 20 kyr is due to the large decrease in summer air temperatures resulting from a large difference in surface albedo between present ice-free summers and ice-age summers, which decreases summer ablation at the ice surface. Ablation at the ice sheet surface remains low because the air temperatures are so low, due to the lapse rate and the higher elevation of the surface, that insolation changes do not increase them enough to warm the surface to the freezing point. As a result the ice sheet maintains itself with a balance of snowfall at its surface, and ablation at its edges. The same seems to be true for B equal to 6 and 8 W m^{-2}K^{-1} although the equilibrium size is much smaller. The situation begins to change for B equal to 9 W m^{-2}K^{-1}. In this case the ice reaches the same maximum size as when B was set to 6 and 8 W m^{-2}K^{-1}; however, when the summer insolation begins to increase at about 11 kyr BP ablation becomes large enough to decrease the ice sheet size appreciably. When B is equal to 10 and 12 W m^{-2}K^{-1} the ice sheet melts away completely within a few thousand years. In this case the air temperatures are so insensitive to insolation variation, due to the large values of B, that the computed net budget is very similar to present-day conditions. As a result ablation increases as the ice sheet shrinks until it melts away completely. In the experiments described below the value of B is set at 8 W m^{-2}K^{-1}, unless otherwise noted.

The energy balance net budget experiments show very little variation in ice sheet cross-sectional area over time compared

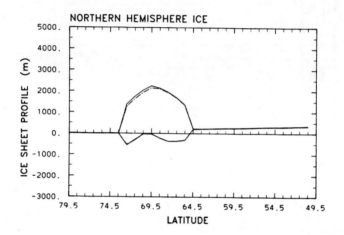

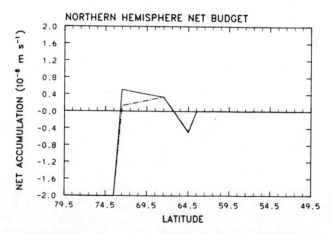

Figure 3 Cross section of northern hemisphere ice sheet and corresponding net budget computed using the energy balance net budget parameterization with A equal to 0.15 bar^{-3}yr^{-1} and B equal to 8 W m^{-2}K^{-1} at -14 kyr (solid line) and -6 kyr (single-dot dashed line). Cross sections correspond to ice areas in Figure 2.

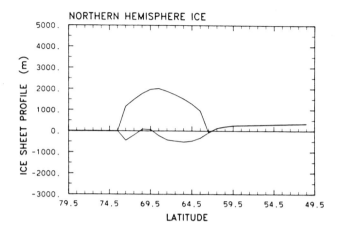

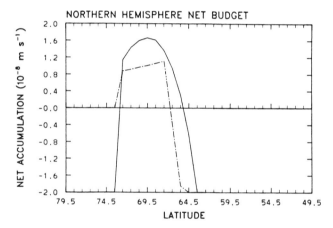

Figure 4 Cross section of northern hemisphere initial ice sheet at −48 kyr and the corresponding net budget as computed by the equilibrium line net budget parameterization (solid line) and the energy balance net budget parameterization (single-dot dashed line).

to that using the equilibrium line net budget parameterization. Figure 3 shows the ice sheet profile and net budget at times of relative extremes of ice sheet size for $A = 0.15 \text{ bar}^{-3} \text{ yr}^{-1}$. The ice sheet does not extend farther south during the ice age period; it simply grows thicker due to increased snowfall at its surface. The reason the ice can not extend farther south is that a large negative net budget is computed at the southern edge of the ice and the ice sheet is too small for flow from the north to make up the mass loss.

The latitudinal position of the southern edge of the ice sheet remains at 63.5°N-62.5°N in most of the experiments. Variation in air temperatures due to the coupling with absorbed insolation variations affects the magnitude of the net budget, but the position of zero net budget, the firn line, is changed very little.

One experiment does show a major change in the positions of the firn line. This is the case in which B equals $4 \text{ W m}^{-2}\text{K}^{-1}$. Here the temperature change is so great that the ice sheet extends to 59.5°N. However, once this extent is reached the position of zero net budget varies very little, and the ice sheet size essentially remains the same.

Figure 4 shows how the net budget for the equilibrium line and energy balance calculations differ for the same ice sheet. The equilibrium line calculation shows a positive net budget over about 75% of the ice sheet compared to about 50% for the energy balance calculation. In addition, the magnitude of the positive net budget in the equilibrium line case is larger than that in the energy balance case. This is due in part to the incorporation of the "desert-elevation" effect (the exponential decrease of water available for snow with an increase in elevation) in the snowfall parameterization used here.

However, the net ablation zone in the energy balance computation is larger, and, more important, the magnitude of the ablation is much larger than for the equilibrium line calculation (-0.63×10^{-8} m s^{-1} and -0.5×10^{-7} m s^{-1} for the energy balance calculation compared to -0.22×10^{-7} m s^{-1} and $+0.95 \times 10^{-8}$ m s^{-1} for the equilibrium line calculation at 63.5°N and 66.5°N respectively). As a result the ice sheet grows much larger using the equilibrium line calculation, while it shrinks back to its "equilibrium" latitudinal extent in a few thousand years using the energy balance calculation.

One possible reason for the extreme ablation over the southern edge of the ice sheet is that the net budget is computed for five-degree latitude zones and then interpolated to the one-degree latitudinal grid of the ice sheet model. This may cause an interpolation from a small positive net budget over ice to a large negative budget over ice-free land, putting the southern edge of the ice into a zone of a large ablation.

In order to determine if removing this large ablation at the southern ice edge would allow the ice sheet to extend farther south during an ice age, an experiment is done in which the net budget is reset to zero in any latitudinal zone in which there is no ice. The interpolation across the southern ice edge is then between a positive net budget over ice and zero. While this modification changes the shape of the ice sheet, it does not allow the ice sheet to extend farther south.

SEA ICE SENSITIVITY EXPERIMENTS

Unlike the continental ice sheet studies, the sea ice experiments are performed at individual latitudes at which the equilibrium seasonal cycle is determined. Long time series of sea ice response are unnecessary in this study because sea ice responds very quickly, on the order of 10 years, to changes in the energy balance compared to the time scale of the changes in insolation, on the order of 10^4 to 10^5 years. As a result, comparisons can be made between different equilibrium experiments as if they were "snap shots" in a time series.

Insolation at -24 kyr, -10 kyr, and the Present

Figure 5a shows the response of the sea ice at 72.5°N and 67.5°S to the insolation regimes at -24 kyr (a period of lower northern hemisphere summer insolation), -10 kyr (a period of higher northern hemisphere summer insolation), and the present (Figure 5b shows these insolation regimes). The results show that sea ice thickness does not vary appreciably with variations in insolation, though there is a slight difference in the period of time ice-free conditions exist. This variation in the period of ice-free ocean ranges from one to two extra weeks in the northern hemisphere, to three less weeks in the southern hemisphere at -10 kyr and -24 kyr compared to the present. This occurs because at -10 kyr, during the period of ice formation, the insolation is the lowest of the three regimes, and as a result the sea ice forms earlier.

Varying the Air Temperature

The sensitivity of the sea ice to changes in the mean and amplitude of the specified seasonal cycle of air temperature was examined by varying these parameters.

Figure 6a shows the response of the sea ice model at 72.5°N and 67.5°S to changes in the mean of air temperature of +5 K, +3 K, 0 K, -3 K, and -5 K (Figure 6b shows the temperature variations). In the northern hemisphere major changes occur in both the ice thickness (0.95 m to 2.5 m maximum ice thickness at 72.5°N) and the period of completely ice-free ocean (6.5 to 0 months at 72.5°N). In the southern hemisphere the sea ice thickness varies to a much greater extent than for insolation

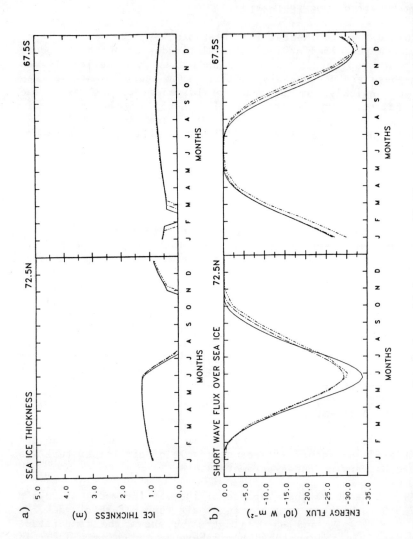

Figure 5 a) Seasonal cycle of sea ice thickness at 72.5°N and 67.5°S for insolation at -10 kyr (solid line), -24 kyr (single-dot dashed line), and the present (double-dot dashed line). b) Seasonal cycle of insolation present at the surface at the corresponding times.

changes alone (0.55 m to 0.95 m at 67.5°S); however, the major change is in the period of ice-free ocean (4.5 to 0 months at 67.5°S).

While changes in the ice thickness are important, especially when ice is thin, changes in the period of open ocean have major consequences on the total energy exchange between the atmosphere and ocean and thus on climate.

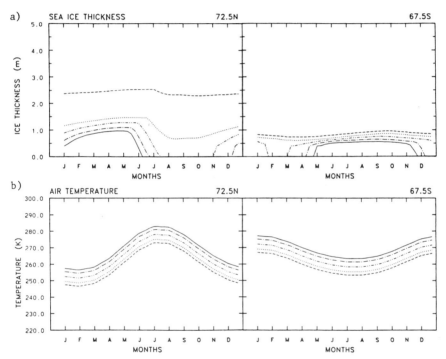

Figure 6 a) Seasonal cycle of sea ice thickness for changes in the mean air temperature of +5 K (solid line), +3 K (single-dot dashed line), 0 K (double-dot dashed line), -3 K (dotted line), and -5 K (dashed line). b) Seasonal cycle of air temperature for the corresponding changes in its mean.

An example of how effective the sea ice can be at insulating the ocean from the atmosphere can be seen by comparing the energy exchanges between the ocean and atmosphere at 72.5°N for the present temperature cycle and when the mean temperature is dropped 3 degrees. Table 1 shows the individual fluxes at the ice-atmosphere interface in early November, when the ocean is still ice-free for present conditions and completely ice-covered for the decreased temperature conditions. The net effect is that the presence of ice 0.9 m thick cuts the net flux from the ocean to the atmosphere by almost a factor of 5, from 123 W m^{-2} to 26 W m^{-2}. When coupled to a climate model this would create a positive feedback which would cool the air further and in turn allow the formation of more ice.

Table 1a. Temperatures of Atmosphere, Ice, and Ocean at the Atmosphere-Ice/ocean Interface in Early November at 72.5°N.

Change in Mean Air Temperature	0.0 K	-3.0 K
Air Temperature	260.2 K	257.2 K
Snow/Ice Interface Temperature	-	252.6 K
Ocean Temperature	271.4 K	271.2 K
Fraction of Open Water	1.00	0.01

Table 1b. Energy Exchanges Between Atmosphere and Ice/Ocean in Early November at 72.5N in W m^{-2}.

Change in Mean Air Temperature	0 K		-3 K	
	ice	ocean	ice	ocean
Short-Wave	-	-1.96	-2.53	-1.96
Albedo	-	0.10	0.84	0.10
Latent Heat	-	8.53	-0.81	9.32
Sensible Heat	-	16.61	-6.84	20.81
Long-Wave from Surface	-	308.	231.	307.
Long-Wave from Atmosphere	-208.	-208.	-196.	-196.
Sum of Fluxes weighted by fraction of open ocean		123.		26.

Note: A positive flux is an upward flux.

Experiments were also performed increasing and decreasing the amplitude of the seasonal cycle of air temperature. Unlike changing the temperature mean, this has little effect on the maximum ice thickness; however, it has a major effect on the period of ice-free conditions. At 72.5°N ice-free conditions last 4.5 months for a 20% increase in the amplitude of the temperature. This period is reduced to 1 month for a similar decrease in the amplitude of the temperature cycle. At 67.5°S the period of ice-free conditions ranges from 2 to 0 months. Thus, varying the amplitude of the seasonal cycle of air temperature has effects similar to those for variations in the mean of the air temperature. Reducing the period of time that ice-free conditions exist reduces the energy exchange with the atmosphere, thus cooling temperatures further, and reducing moisture available for snowfall.

SUMMARY AND CONCLUSIONS

This work confirms the work of others that suggests that cryospheric variations on the order of 100 kyr can be generated using zonally averaged ice sheet models. When an equilibrium line net budget parameterization is used this continental ice sheet model successfully reproduces the slow buildup of ice from 125 kyr BP to 18 kyr BP, including a maximum ice sheet extent at 18 kyr BP, a large deglaciation from 18 kyr to 6 kyr BP, and shorter-term variations on the order of 20 kyr to 40 kyr. However, these cryospheric variations were not seen and it was difficult to initiate an ice sheet using an energy balance net budget calculation in place of the equilibrium line calculation. This is despite the fact that the energy balance calculation should more closely approximate the physical processes involved in determining the net budget than the equilibrium line parameterization. Thus, either the strategy employed for calculating net budget using an energy balance approach is inadequate to model accumulation and ablation in zonally averaged ice sheet models, or zonally averaged ice sheet models themselves may be inadequate to translate the variations calculated from an energy balance net budget and changing solar radiation into fluctuations on a climatic scale.

The energy balance net budget calculation and the equilibrium line parameterization behave very differently at the firn line. The variation in the latitudinal position of the firn line is much greater for the equilibrium line calculation than for the energy balance calculation in response to equivalent fluctuations in solar radiation. In part the discrepancy between the results of the energy-balanced and equilibrium line calculations may be related to the method by which air temperature is coupled to insolation variations. In the energy balance strategy latitudinally and seasonally varying planetary albedo is specifically included in the calculation of surface air temperature, while in the equilibrium line parameterization, albedo is simply accounted for in a constant coefficient.

The seasonal variation in albedo can be very important in determining whether or not snow can be maintained through the summer and therefore allow an ice sheet to grow. In regions where there is no ice the albedo is relatively low, the surface absorbs a large percentage of the short-wave radiation, and therefore the snow melts away in the summer season. If it were possible to maintain the snow through the summer either through large ice flow into an ice-free region from an existing ice sheet or through regional effects such as snow accumulation on high-altitude plateaus and on land areas which lie along natural storm tracks, the ice sheet may be more responsive to fluctuations in insolation and temperature as they are reflected in the energy balance net budget. However, in this model the ice sheet is not large enough to flow past $\sim 62.5°N$, and regional effects are not taken into account.

There are several other aspects of a complete net budget calculation that are not accounted for in these models. One critical feature that is neglected is the contribution of the oceans and sea ice to snow accumulation over land surfaces. In an isolated ice sheet model continental ice is decoupled from its moisture source, namely the ocean. The extent of sea ice controls the availability of moisture for snowfall. As the sea ice extends farther south during an ice age it caps off the oceans and thus reduces the available moisture. Conversely, the retreat of sea ice should have the opposite effect. Thus coupling sea ice and ice sheet fluctuations may be important for calculating a more realistic net budget.

An isolated sea ice model has been used in experiments to test its response to variations in insolation and temperature as a first step toward the coupling of sea ice-ice sheet fluctuations. These experiments show that the thickness and extent of sea ice, while relatively insensitive to insolation changes alone, are quite sensitive to changes in air temperature. Although the changes in the sea ice thickness can be important in regulating the energy exchange between the atmosphere and ocean, the changes in the period of ice-free conditions have a dramatic effect on that exchange. Variations in summer air temperature on the order of $+3 K$ to $-3 K$ can change the period that ice-free conditions exist from six to zero months at $72.5°N$. The presence of ice in previously ice-free zones can cause a factor-of-five decrease in the net flux of energy from the ocean to the atmosphere. Such a decrease can induce further cooling of air temperatures and thus the formation of more sea ice. These results indicate that the variations in sea ice in response to changes in air temperatures on climatic time scales may have important implications for the net budget over land areas and for further variations in air temperature. The sensitivity of the air temperature and snowfall to changes in the sea ice extent will be better understood once the sea ice is coupled to a climate model. However, such a coupling is premature. Further studies of the sensitivity of the sea ice to the numerous parameterizations and simplifications of the complex physical processes important in sea ice growth, such as the lead parameterization and the use of a simple mixed-layer ocean, are needed before the results of such a coupled model could be meaningfully interpreted.

REFERENCES

1. Berger, A.L. : 1978, J. Atmos. Sci. 35, pp. 2362-2367.
2. Thompson, S.L. : 1979, in : "Report of the JOC Study Conference on Climate Models : Performance, Intercomparison, and Sensitivity Studies", vol. II, W.L. Gates (Ed.), pp. 1002-1023, GARP Publication Series 22.
3. Thompson, S.L., and Barron, E.J. : 1981, J. of Geology 89, pp. 143-167.
4. Semtner, A.J. : 1976, J. Phys. Oceanog. 6, pp. 379-389.
5. Maykut, G.A. : 1978, J. Geophys. Res. 83 C7, pp. 3646-3658.
6. Paterson, W.S.B. : 1969, "The Physics of Glaciers", Pergamon Press, New York, 250pp.
7. Glen, J.W. : 1955, Proceedings of the Royal Society of London 288, pp. 519-538.
8. Pollard, D. : 1982, Nature 296, pp. 334-338.
9. Birchfield, G.E., Weertman, J., and Lunde, A.T. : 1981, Quaternary Research 15, pp. 126-142.
10. Ledley, T.S. : 1983, A Study of Climate Sensitivity Using Energy Balance Cryospheric Models, Ph.D. dissertation, Massachusetts Institute of Technology, 314 pp.
11. Warren, S.G., and Schneider, S.H. : 1979, J. Atmos. Sci. 36, pp. 1377-1391.
12. North, G., Mengel, J.G., and Short, D.A. : 1984, in : "Milankovitch and Climate", A. Berger, J. Imbrie, J. Hays, G. Kukla, B. Saltzman (Eds), Reidel Publ. Company, Holland. This volume, p. 513.

[1]Present address : Department of Space Physics and Astronomy, P.O. Box 1892, Houston, TX 77251, USA.

A POSSIBLE EXPLANATION OF DIFFERENCES BETWEEN PRE- AND POST-JARAMILLO ICE SHEET GROWTH

R.G. Watts[1], Md. E. Hayder[2]

[1] Department of Mechanical Engineering, Tulane University, New Orleans, Louisiana, USA

[2] Department of Aerospace and Mechanical Engineering, Princeton University, Princeton, New Jersey, USA

ABSTRACT

The response of a simple ice sheet model forced by a periodic change in the climate point exhibits a bifurcation when certain dimensionless parameters undergo realistic changes. These dimensionless parameters are related to the poleward limit of continents as well as to the frequency of the climate point variation and other parameters. It is postulated that this bifurcation may be responsible for the initiation of large glacial cycles about 900 000 years ago following smaller, higher frequency ice sheet advances and retreats before that time.

Data from deep sea cores indicate that large scale climatic shifts occurred at 1.3 m.y. BP and 0.8 m.y. BP (1). Beginning about 1.3 m.y. BP the climate was punctuated by short, cold pulses. The durations of these cold pulses appears never to have exceeded 30 000 years. Following 0.9 m.y. BP both the extent and duration of these cold periods increased rather suddenly as the earth entered the Pleistocene ice age. We suggest here that the transition from small, relatively high frequency ice sheet advances to the more recent larger, longer period ice sheet advances could have been triggered by a minor change in any of several parameters that determine the ice sheet response to orbital parameter variations.

We begin by writing down the differential equation describing the plastic flow of a two dimensional ice sheet following Weertman (2).

$$\frac{2\sqrt{\gamma}}{3}\frac{d}{dt}[(1+\eta)L^{3/2}] = A \qquad [1]$$

where γ is a constant estimated by Weertman (2) to be ~7-14m, A is the net accumulation rate on the equatorward half of the ice sheet, L is the half length of the ice sheet, t is time and η is the ratio of the depth of the bedrock below its position with no ice sheet present to the height above the position with no ice sheet present, h_B/h, assumed to be a function of time only. Following Oerlemans (3) we assume that bedrock sinking can be modeled as a first order lag with time constant τ_B, i.e.,

$$\frac{dh_B}{dt} = \frac{1}{\tau_B}(\frac{h}{3} - h_B) \qquad [2]$$

We use the following accumulation rate, which is similar to those used by Weertman (2) and Oerlemans (3):

$$A = a[H-S(2L-P)]L - b[H-S(2L-P)]^2 L, \quad H-S(2L-P) < \frac{a}{2b}$$
$$A = \frac{a^2 L}{4b}, \quad H-S(2L-P) \geq \frac{a}{2b} \qquad [3]$$

The variables are defined in the legend of Figure 1.

Past changes in the earth's orbital elements have caused quasiperiodic changes in the distribution of solar radiation at the top of the atmosphere. Hayder (4) studied the response of a seasonal climate model to these changes and concluded that the change in the location of the climate point P (see Fig. 1 legend) also varies quasiperiodically. Although it is a considerable oversimplification to assume that P varies with a single characteristic period, we might reasonably hope to learn something of the basic nature of the system by studying its response to periodic signals at one or more frequencies (5). We write, therefore,

$$P(t) = P_0 + \Delta P \sin(2\pi t/T) \qquad [4]$$

where T is the period.

When the dimensionless parameters

$$R^2 = 4S^2 L/\gamma \qquad p = 2S^2 P/\gamma \qquad \tau = at/2$$
$$\kappa = b\gamma/2aS \qquad \mu = 2/a\tau_B \qquad \Gamma = \eta R$$
$$\beta_0 = 2S^2 P_0/\gamma \qquad \beta_1 = 2S^2 \Delta P/\gamma \qquad \omega = 4\pi/aT \qquad [5]$$

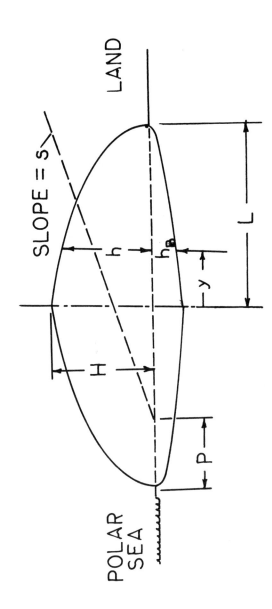

Figure 1 Ice sheet under plastic flow. 2L is the latitudinal extent of the ice sheet measured from the poleward limit of the continent. P is the distance from the poleward limit of the land to the point where sub-freezing temperatures exist at the sea level year round. Equatorward of P the altitude of the freezing point increases with slope s. $H = (\gamma L)^{1/2}$ is the centerline height of the ice sheet. a and b are constants.

are substituted into eqs 1-4 the following equations result :

$$\frac{1}{3R^2}\frac{d}{d\tau}\left[(1+\frac{\Gamma}{R})R^3\right] = (R-R^2+p)[1-\kappa(R-R^2+p)], R-R^2+p < \frac{1}{\kappa} \quad [6]$$

$$= \frac{1}{4\kappa}, \quad R-R^2+p \geq \frac{1}{2\kappa}$$

$$\frac{1}{\mu}\frac{d\Gamma}{dt} = \frac{R}{3} - \Gamma \quad [7]$$

$$p = \beta_0 + \beta_1 \sin(\omega\tau) \quad [8]$$

We have studied the response of the system described by eqs 6,7, and 8 for various values of the dimensionless parameters. A more extensive paper describing our results is in preparation. We report here a result which we believe is of considerable interest.

The parameters used above have the following orders of magnitude (see, e.g., Oerlemans (3)) ; $s=10^{-1}$, $\gamma=7m$, $P_0=-150km$, $\Delta P=300km$, $a=0.73 \times 10^{-3} yr^{-1}$, $b=0.27 \times 10^{-6} m^{-1} yr^{-1}$, $T=20\,000$ yrs, $\tau_B=5\,000$ yrs. When these values are used in eq. 5, we find that $\beta_0=-0.0429$, $\beta_1=0.0714$, $\kappa=1.29$, $\omega=0.938$, and $\mu=0.55$. The response of R (the square root of dimensionless ice sheet length) is shown in Figure 2 for parameter values corresponding approximately to these, but for a small range to β_0 values. An infinitesimal change in β_0 at some value between -0.0427 and -0.0426 causes the solution to bifurcate, i.e., to change behavior qualitatively. When $\beta_0<-0.0427$ a series of small ice sheets alternately grow and disappear. However, when $\beta_0>-0.0426$, the ice sheet size fluctuates somewhat but growth continues until the size begins to oscillate around some large value. In order for the ice sheet to collapse, some new mechanism must be added. Andrews (6) and Pollard (7) suggested that ice sheets might rapidly disappear due to calving into the ocean or proglacial lakes once they become very large. After this occurred, the cycle would begin again. Since it would take several periods for the ice sheet to become large enough for calving to occur, the period would be large.

It has already been pointed out by Birchfield (5) that small changes in the frequency of the forcing function (P(t)) can result in radical changes in ice sheet behavior. We point out here that similar behavior can result from small changes in any of the terms in the dimensionless parameter β_0 as well. The changes can be infinitesimal.

DIFFERENCES BETWEEN PRE- AND POST-JARAMILLO ICE SHEET GROWTH

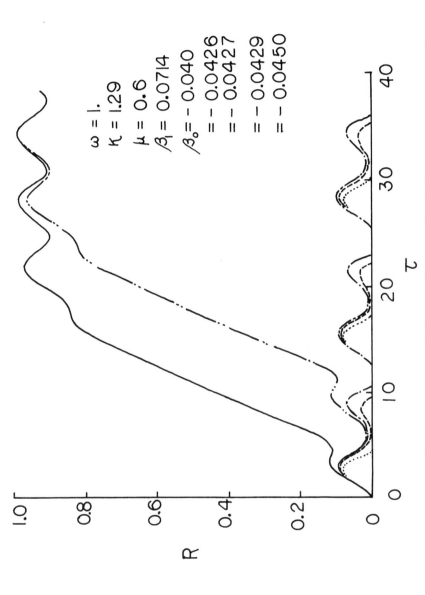

Figure 2 Response of the ice sheet to periodic variations in P(t). R = 1 corresponds to an ice sheet length of 2L ~ 3500 km.

The fact that such small (theoretically infinitesimal) changes in certain parameters can cause very large changes in the ice sheet response is of considerable interest. Specifically, it may explain the sudden transition reported by Ruddiman between pre- and post-Jaramillo glacial cycles. Any one of several physical phenomena could have caused the transition. A decrease in the magnitude of β_0 could result, for example, if the poleward edge of a continent drifted closer to the pole. It could also be brought about by a general decrease in global temperature, an increase of altitude near the poleward edge of the continent (mountain building), or by a decrease in s, which is at least partly controlled by the poleward transport of heat.

Finally, we wish to emphasize our major point : the differential equation that is often used to describe the dynamics of ice sheet growth in response to solar orbital variations bifurcates under conditions of reasonable parameter values and variations. While no particular physical mechanism can be identified (although several suggest themselves), the bifurcation phenomena itself might help in understanding our entry into the present glacial age.

REFERENCES

1. Ruddiman, W.F. : 1971, Geol. Soc. of Amer., Bull. 82, pp. 283-302.
2. Weertman, J. : 1976, Nature 261, pp. 17-20.
3. Oerlemans, J. : 1980, Nature 287, 430-432.
4. Hayder, M. : 1982, M.S. Thesis in Mechanical Engineering, Tulane University.
5. Birchfield, G.E. : 1977, J. of Geophys. Res. 82, pp. 4909-4913. Birchfield has used just such an approach with a model similar to the one presented here. In fact, he discovered that under certain conditions, small changes in the frequency of P(t) can lead to large changes in ice sheet behavior. We show here that small changes in other parameters can also lead to radical changes ice sheet behavior.
6. Andrews, J.T. : 1973, Arct. Alp. Res. 5, pp. 185-199.
7. Pollard, D. : 1983, Nature 296, pp. 334-338.
8. Supported by grant ATM 79-16332 from the National Science Foundation.

ON LONG PERIOD INTERNAL OSCILLATIONS IN A SIMPLE ICE SHEET MODEL

G. E. Birchfield[1], J. Weertman[2]

[1] Department of Geological Sciences and Department of Engineering Sciences and Applied Mathematics, Northwestern University

[2] Department of Geological Sciences and Department of Materials Science and Engineering, Northwestern University, Evanston, Ill. 60201

ABSTRACT

Some numerical experiments with a power law rheology ice sheet model were presented. Bedrock deformation under varying ice loading was incorporated in the simplest possible way.

The focus of the model study was on the question as to whether such simple physics could exhibit self-sustained oscillations of long period (see also 1,2,3). The relevance of such physics is to the prominence of the 100 kyr component in the spectrum of proxy ice volume time series. Such oscillations have been correlated with the eccentricity variations of the earth's orbit; but because of the very small amplitude of the resulting insolation perturbations, it has been difficult to be able to associate them causally (4,5).

The experiments were designed with a snow line fixed in time; starting from no ice sheet the model experiments ran for a few hundred thousand years. In a number of experiments instead of approaching a steady state, the ice volume approached an oscillatory state with a period ranging from 60 to 80 kyr. The amplitude was of the order of 15 to 20 % of the mean ice volume. Repeated experiments indicated that the oscillations exist only for a small range of model parameters. The phenomenon appeared only for very large ice sheets; also needed was a large horizontal gradient in the net accumulation at the surface of the ice sheet.

Approximate analytical solutions of perfect plasticity physics in the same context, appeared to show only strongly damped oscillations to be possible. Further analytical studies of the power law rheology model, made after the Symposium, also failed to indicate the presence of such oscillations.

On close examination of the numerical method of solving the highly nonlinear ice sheet equations, several difficulties became apparent. The model has been reprogrammed with what appears to be a more accurate algorithm. With a large number of calculations for a wide range of model parameters, no evidence of the oscillations has appeared. It is concluded that the earlier model response represented an example of a numerical nonlinear instability and, that it is most likely that such a simple physical system does not have the capability for self-sustained oscillations.

REFERENCES

1. Buys, M. and Ghil, M. : 1984, in : "Milankovitch and Climate", A. Berger, J. Imbrie, J. Hays, G. Kukla, B. Saltzman (Eds), Reidel Publ. Company, Holland. This volume, p. 55.
2. Peltier, W.R. and Hyde, W. : 1984, in : "Milankovitch and Climate", A. Berger, J. Imbrie, J. Hays, G. Kukla, B. Saltzman (Eds), Reidel Publ. Company, Holland. This volume, p. 565.
3. Nicolis, C : 1984, in : "Milankovitch and Climate", A. Berger, J. Imbrie, J. Hays, G. Kukla, B. Saltzman (Eds), Reidel Publ. Company, Holland. This volume, p. 637.
4. Hays, J.D., Imbrie, J. and Shackleton, N.J. : 1976, Science 194, pp. 1121-1132.
5. Imbrie, J., Hays, J.D., Martinson, D.G., McIntyre, A., Mix, A.C., Morley, J.J., Pisias, N.G., Prell, W.L., and Shackleton, N.J. : 1984, in : "Milankovitch and Climate", A. Berger, J. Imbrie, J. Hays, G. Kukla, B. Saltzman (Eds), Reidel Publ. Company, Holland. This volume, p. 269.

ON THE ORIGIN OF THE ICE AGES

J. Oerlemans

Institute of Meteorology and Oceanography, University of Utrecht, Utrecht, The Netherlands

ABSTRACT

Ice sheet dynamics provide a possible explanation for the 100 kyr power in climatic records. Some numerical experiments presented here show that even the transition from an essentially ice-free earth to a glacial cycle regime can be produced by a northern hemisphere ice-sheet model, provided that a slow general cooling on the northern hemisphere continents is imposed. Such a cooling could for example be the result of continental drift.

INTRODUCTION

Now that the Pleistocene glacial cycles have been documented quite well, e.g. Berger (1), more and more attention is paid to model simulation of the ice-volume record. Recent experiments (1,2,3,4) have shown that power on the 100 kyr time scale may be almost entirely due to ice-sheet dynamics. Here destabilization of an ice sheet by bedrock sinking and accumulation of heat plays an important role. Results from model experiments with both effects included were recently reported by the author (5).

Accepting that the regularity (on the 100 kyr time scale) of the ice ages is indeed mainly due to the physical properties of ice sheets, the question 'How did the Pleistocene start' can be considered. According to for instance Shackleton and Opdyke (6) and Hooghiemstra (7) the climatic evolution to the Pleistocene was rather discontinuous. The transition from an essentially ice-free climatic regime to one with regular glacial cycles occurred in a relatively short time. In this contribution some arguments will be given showing that such a rapid transition is in agreement with the explanation of glacial cycles from ice-sheet dynamics.

Results shown here were all obtained with the northern hemisphere ice-sheet model described in Oerlemans (5). Readers are referred to this article for a model description and general discussion.

BASIC IDEA

An ice sheet may form on the northern hemisphere continents when the snow line intersects the surface south of the polar sea, see Figure 1. Since the snow-line height will vary (due to, for instance, the Milankovitch insolation variations), the ice accumulation on the continent will be positive during a limited time. From model studies quoted above, we know that once an ice sheet has reached some size, it is able to continue its growth through the height-mass balance feedback, even when the snow line goes up again. So the probability that a large ice sheet can be established increases with the strength of the forcing (up and down movement of the snow line), as well as the mean elevation of the snow line.

As is immediately clear from Figure 1, a northward shift of the northern hemisphere continents is equivalent to a lowering of the (mean) snow line. Together with the fact that long-period internal oscillations may occur once the ice sheet has reached some critical size (5), it is obvious that a gradual northward shift of the northern hemisphere continents may lead to a rapid transition from an almost ice free northern hemisphere to a climate with glacial cycles. Precisely the same argument applies to every gradual global climatic cooling, whatever its cause may be.

From these considerations it is clear that, although the transition from one to another climatic regime may be well defined, the point in time at which the transition occurs is definitely not.

The transition is more likely to occur when variability of the snow-line elevation is large, and therefore a period with large eccentricity of the earth's orbit seems to be favourable.

However, movement of the snow line will have a substantial stochastic component, making the point(s) in time at which transition occurs practically unpredictable.

ON THE ORIGIN OF THE ICE AGES

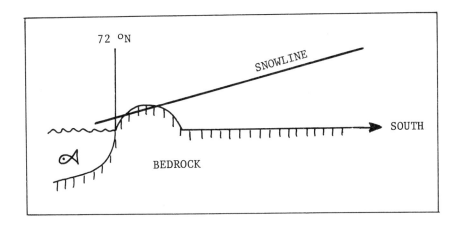

Figure 1 Northern hemisphere geometry for the ice-sheet model. The topographic bump is introduced to represent the plateaus that are actually found at high latitudes.

SOME MODEL CALCULATIONS

The ice-sheet model of which results will be discussed now is based on the geometry shown in Figure 1. It calculates ice flow along a meridian and includes a computation of the temperature field in the ice sheet, thus giving the possibility to take into account the effect of ice temperature and basal-water formation on the ice-mass discharge. The time-dependent reaction of the bedrock to ice loading is also calculated.

In Figure 2, two model runs are shown covering 1 million years of simulated time. In both runs the elevation of the snow line was prescribed to decrease steadily by 1 m per 5 kyr, while in the second run additional random forcing (white noise) was added. Model constants were chosen in such a way that transition occurs in the simulated range of time.

Run 1 shows very clearly how, at some critical value of the snow-line elevation, a self-sustained internal oscillation is set up. The oscillation is of the relaxation type : a gradual building-up of the ice sheet followed by a rapid decay (in accordance with the observational evidence). The small amount of ice present before the transition to the glacial cycle regime reflects some minor ice cover on the mountain (Fig. 1).

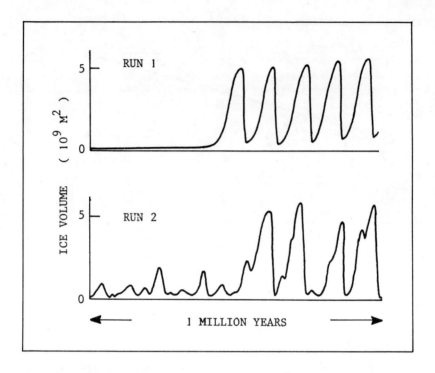

Figure 2 Two model runs of 1 million years of simulated time. The snow line elevation decreases steadily. In run 2 a random component is added to the movement of the snow line.

According to run 2, additional forcing on a smaller time scale makes the result look more realistic. Still a distinct change in the character of the solution occurs (for other "global" model constants), but the transition is smoothed somewhat by the presence of response in smaller time scales. Additional experiments confirmed that transition to a 100 kyr glacial cycle regime does not depend on the precise form of the forcing, as long as it does not have a strong component with a time scale close to or larger than the internal period of the ice sheet.

In conclusion, although the Milankovitch insolation variations may steer the Pleistocene glacial cycles, it appears that an ice-sheet model forced by random forcing of sufficient strength is capable of producing a transition from a comparatively ice-free to a glacial cycle climatic regime. A very slow, cooling of the global climate (associated with continental drift, for example) can easily cause such a transition.

REFERENCES

1. Berger, A. : 1981, Climatic Variations and Variability : Facts and Theories, 795pp., Reidel Publ. Company, Holland.
2. Oerlemans, J. : 1980, Nature 287, pp. 430-432.
3. Birchfield, G.E., Weertman, J., and Lunde, A.T. : 1981, Quaternary Research 15, pp. 126-142.
4. Pollard, D. : 1982, Nature 292, pp. 233-235.
5. Oerlemans, J. : 1982, Climatic Change 4, pp. 353-374.
6. Shackleton, N.J., and Opdyke, N.D. : 1976, Geol. Soc. Am. Mem. 145, pp. 449-464.
7. Hooghiemstra, H. : 1984, in : "Milankovitch and Climate", A. Berger, J. Imbrie, J. Hays, G. Kukla, B. Saltzman (Eds), Reidel Publ. Company, Holland. This volume, p. 371.

PART III

MODELLING LONG-TERM CLIMATIC VARIATIONS IN RESPONSE TO ASTRONOMICAL FORCING

SECTION 3 – OSCILLATOR MODELS OF CLIMATE

EARTH-ORBITAL ECCENTRICITY VARIATIONS AND CLIMATIC CHANGE

B. Saltzman[1], A. Sutera[2], A.R. Hansen[1]

[1]Department of Geology and Geophysics, Yale University, New Haven, CT 06511, USA

[2]The Center for the Environment and Man, Inc., Hartford, CT 06120, USA

ABSTRACT

The possible role of direct long-period forcing of the climatic system due to earth orbital eccentricity variations of 100 kyr period is explored.

The model treated involves marine ice shelf extent and bulk ocean temperature as coupled prognostic variables, and can be specialized to admit either an exponentially stable equilibrium or an unstable equilibrium near which stable limit cycles of arbitrary period are possible. The conditions under which the eccentricity variations can force ice variations of the observed magnitude are determined. The responses obtained can be realistic, but for reasonable ranges of the parameters are found to be structurally unstable. The role of random perturbations on the responses are also examined.

INTRODUCTION

In recent studies we showed that with simple approximations involving marine ice shelf extent and bulk ocean temperature as coupled prognostic variables it is possible to construct a plausible climatic dynamical model exhibiting free auto-oscillations of both short and long periods (1,2). We now address several further questions that come to the fore regarding the influence of periodic forcing on such a system, e.g.,

1) Within the framework of the model treated, can long-period oscillations arise simply by the direct forcing known to exist at approximately 100 kyr due to the earth-orbital eccentricity variations discussed by Milankovitch (3) (cf.(4)) ?

2) What will be the consequence of the interaction of this known "eccentricity forcing" with the free oscillatory solutions obtained previously? In particular, to what degree can the eccentricity forcing affect the free oscillatory response of a nonlinear climatic system, and, conversely, to what degree can the nonlinearities associated with free oscillations affect the long-period forced response?

3) What is the effect of aperiodic fluctuations (i.e., white noise) on the responses?

The physical model to be considered is the one introduced in (5) based on the mass balance of marine ice and the thermal balance of the entire ocean. This system can serve as a prototype of other two-component climatic systems that admit a range of linear and nonlinear responses including limit cycle behavior of an arbitrary frequency ω_L. A brief review of the model, including specifications of the flux parameterizations, eccentricity forcing, and noise, is given in the next section. At this time we do not consider the role of cyclic forcing of periods near 20 kyr and 40 kyr, due to precessional and obliquity variations, that must also be operative (cf.,(6)).

THE MODEL

As we have noted, we shall study the system described in (5) based on the mass balance of marine ice and the thermal balance of the entire ocean. This system is governed by a coupled pair of equations for the time variations of the 10-year running average values of the sine of the ice-edge latitude η, and the mean ocean temperature θ, respectively, i.e.,

$$\frac{\delta \eta'}{\delta t} = \Phi + R_{\eta 1} \qquad [1]$$

$$\frac{\delta \theta'}{\delta t} = \Psi + R_{\theta 1} \qquad [2]$$

If we assume that continental ice mass is diagnostically related to the marine ice mass, η can be taken as a measure of the total ice mass of the planet.

In [1] and [2] the primes denote departures from an equilibrium at $(\eta', \theta')=(0,0)$, R_1 represents stochastic forcing arising from the impossibility of fulfilling the Reynolds condition for time derivatives averaged over aperiodic high frequency variations,

$$\Phi = \Phi(\eta',\theta';e) = \frac{1}{\rho_i L_f (2I/\Lambda)} \sum_{n=1}^{5} [H^{(n)\downarrow}(\eta)]'$$

and

$$\Psi = \Psi(\eta',\theta';e) = \frac{1}{\rho_w c_w D} \{ \sum_{n=1}^{4} \widetilde{H}_{sw}^{(n)\downarrow}]' - [\widetilde{H}_{swi}^{(3)\downarrow}]' - L_f \rho_i I \frac{d\eta'}{dt} \}$$

where ρ is density, c is specific heat, L_f is the latent heat of fusion, D is the ocean depth, I is the average ice thickness, $(2I/\Lambda)$ is an "inertia" factor that measures the effect of heat flux imbalances at the ice edge on the net change in ice mass, $H_s^{(n)\downarrow}$ is the downward heat flux at the surface due to short wave radiation (n=1), long wave radiation (n=2), small scale convection (n=3), water phase changes (n=4), subsurface oceanic convection (n=5), subscripts w and i denote the ocean and ice, respectively, and the wavy bar ($\sim$) denotes an integral with respect to the sine of the latitude over all latitudes consisting of either water or ice as denoted by the subscript, and e denotes eccentricity.

We note that the downward flux of solar radiation at the earth's surface, $H^{(1)\downarrow}$ depends on the value of the solar constant, which in turn depends on periodic variations of the eccentricity e. Thus, if we set

$$e = e_0 + e'(t),$$

where e_0 is the very long term (e.g., 10^6 year) mean value of e, we can expand $[H^{(1)\downarrow}]'$ in the form,

$$[H^{(1)\downarrow}(e)]' = [H^{(1)\downarrow}(e_0)]' + H_s^{(1)\downarrow}(e')$$

assuming that the equilibrium at $(\eta',\theta') = (0,0)$ corresponds to $e = e_0$.

Since the eccentricity varies from a minimum of 0.00 (circular orbit) to a maximum of about 0.06, we can set $e_0 \approx 0.03$ (the present value is about 0.02). This range of values of e has been estimated to correspond to a solar constant change of about 2 Wm^{-2} or about 0.2% of its present value of 1340 Wm^{-2} (cf.,(7)). Thus, we can write

$$\Phi = \Phi(\eta', \theta'; e_0) + \frac{1}{\rho_i L_f (2I/\Lambda)} H_{sn}^{(1)\downarrow}(e')$$

$$\Psi = \Psi(\eta', \theta'; e_0) + \frac{1}{\rho_w c_w D} \tilde{H}_{sw}^{(1)\downarrow}(e')$$

where

$$H^{(1)\downarrow}(e') = a \cos \omega_e t$$

$$\omega_e = 2\pi/10^5 y$$

$$a = 0.001 \, H_S^{(1)\downarrow}(e_0)$$

[1] and [2] then become

$$\frac{\delta \eta'}{\delta t} = \Phi(\eta', \theta'; e_0) + A_\eta \cos \omega_e t + R_{\eta 1}(t) \qquad [3]$$

$$\frac{\delta \theta'}{\delta t} = \Psi(\eta', \theta'; e_0) + A_\theta \cos \omega_e t + R_{\theta 1}(t) \qquad [4]$$

where $A_\eta = a_\eta [\rho_i L_f (2I/\Lambda)]^{-1} \approx 10^{-3} H_{sn}^{(1)\downarrow}(e_0)[\rho_i L_f (2I/\Lambda)]^{-1}$, and
$A_\theta = \tilde{a}[\rho_w c_w D]^{-1} \approx 10^{-3} \tilde{H}^{(1)\downarrow}(e_0)[\rho_w c_w D]^{-1}$

In this study we shall adopt the form for $\Phi(\eta', \theta'; e_0)$ given in (1), i.e.,

$$\Phi(\eta', \theta'; e_0) = \phi_1 \theta' - \frac{1}{\tau_\eta} \eta' + R_{\eta 2}$$

where ϕ_1 and τ_η ($=\phi_2^{-1}$, the ice-edge relaxation time) are positive constants and $R_{\eta 2}$ is a random departure from the deterministic part. For $\Psi(\eta', \theta'; e_0)$ we adopt two forms: (I) a simple linear relation, representing relaxation to a stable equilibrium, and (II) the nonlinear relation given in (1), i.e.

$$\Psi(\eta', \theta'; e_0) = \begin{cases} -\dfrac{1}{\tau_\theta} \theta' + R_{\theta 2} & \text{(linear case, I)} \\ -\psi_1 \eta' + \psi_2 \theta' - \psi_3 \eta'^2 \theta' + R_{\theta 2} & \text{(SSE case, II)} \end{cases}$$

where ψ_1, ψ_2, ψ_3 and τ_θ (thermal relaxation time for the entire ocean) are positive constants and $R_{\theta 2}$ is a random departure. These two cases correspond to scenarios in which the climate can fluctuate deterministically near stable and unstable equilibria respectively. It is our goal to examine the role of long-period eccentricity forcing in each of these cases.

From (8, Fig. 5) we find that for a reasonable value of η, $H_{sn}^{(1)} \approx 50$ W m^{-2} and $H_{sw}^{(1)} \approx 160$ W m^{-2}, so that $a_\eta \approx 0.05$ W m^{-2} and $\tilde{a} = 0.16$ W m^{-2}. Taking the values $\rho_i \approx \rho_w = 10^3$ kg m^{-3}, $L_f = 3.34 \times 10^5$ J Kg^{-1}, $C_w = 4.186 \times 10^3$ J kg^{-1} K^{-1}, D=4 km, and assuming $(2I/\Lambda)=10^5$ m corresponds roughly to shelf ice inertia, we obtain,

$$A_\eta \approx 10^{-15} \text{ s}^{-1} \approx 3 \times 10^{-8} \text{ y}^{-1}$$

$$A_\theta \approx 10^{-11} \text{ K s}^{-1} \approx 3 \times 10^{-4} \text{ K y}^{-1}$$

Whereas the amplitude of the eccentricity-induced fluctuations of η due to direct solar radiation variations on the ice edge (measured by A_η) are negligibly small relative to the observed variations of η, it is not similarly obvious that the effects of these solar radiation variations on the mean ocean temperature (measured by A_θ) are too small to be important. It is these latter effects of the eccentricity cycle that we shall consider in this study.

We next estimate the magnitude of the stochastic forcing of η and θ respectively, $R_\eta = R_{\eta 1} + R_{\eta 2}$ and $R_\theta = R_{\theta 1} + R_{\theta 2}$. These quantities can be written in the form

$$R_\eta = \frac{\varepsilon_\eta^{1/2}}{\delta t} \delta W \equiv |R_\eta| \cdot N$$

$$R_\theta = \frac{\varepsilon_\theta^{1/2}}{\delta t} \delta W \equiv |R_\theta| \cdot N$$

representing a white-noise process in which $\delta W \equiv \delta t^{1/2} \cdot N$ can be identified with the output (at each time step, δt) of a Gaussian random number generator N of zero mean and variance unity, and the standard deviation of the random forcing $|R_x| \equiv \sigma(R_x) = (\varepsilon_x/\delta t)^{1/2}$ is the "stochastic amplitude" or "noise level".

A value of $|R_\eta|$ appropriate for random shelf ice variations should be considerably shorter than the value $|R_\eta| = 10^{-3} \text{y}^{-1}$ estimated in (1) for pack ice excursions. A value of 10^{-4}y^{-1} would appear to be reasonable.

An appropriate value of $|R_\theta|$ should correspond to random fluctuations in the surface heat flux of the order of at least 1 W m^{-2}. For our 4 km deep ocean this is equivalent to $|R_\theta| \approx$

3×10^{-3} K y^{-1} (or $\varepsilon_\theta^{1/2} = 3 \times 10^{-3}$ K y$^{-1/2}$ since $\delta t = 1$ year), which we shall take as our reference value of the stochastic noise.

To complete the specifications of the model we must assign values to the constants ϕ_1, τ_η, τ_θ, ψ_1, ψ_2, ψ_3. This will be done in the next two sections in which we discuss the linear case I, and the nonlinear case II, respectively.

CASE I : LINEARLY DAMPED FEEDBACK

We consider the system

$$\frac{\delta \eta'}{\delta t} = \phi_1 \theta' - \frac{1}{\tau_\eta} \eta' + A_\eta \cos \omega_e t + R_\eta \qquad [5]$$

$$\frac{\delta \theta'}{\delta t} = -\frac{1}{\tau_\theta} \theta' + A_\theta \cos \omega_e t + R_\theta \qquad [6]$$

In the absence of stochastic forcing ($R_\eta = R_\theta = 0$), it is easy to show that if τ_θ and τ_η are much smaller than the period of the forcing, $P_e = 2\pi/\omega_e = 100$ kyr, the system is diagnostic and the deterministic solution can be expressed the simple form

$$\theta' = |\theta'|_e \cos \omega_e t \qquad [7]$$

$$\eta' = |\eta'|_e \cos \omega_e t \qquad [8]$$

where

$$|\theta'|_e = \tau_\theta A_\theta \qquad [9]$$

$$|\eta'|_e = (A_\eta + \phi_1 \tau_\theta A_\theta) \qquad [10]$$

From [10] we can see that, although the direct eccentricity forcing of ice extent measured by A_η might be small, this forcing is linearly amplified by the indirect effect of the oceanic interaction measured by ($\phi_1 \tau_\theta A_\theta$). Since τ_θ, τ_η, A_θ, and A_η can all be assigned values within rough observationally-derived limits we can use [9] to determine a range of amplitudes of the oceanic temperature change to be expected if the eccentricity forcing were acting alone without noise, and use [10] to determine a range of values of ϕ_1 ($\approx \partial \dot{\eta} / \partial \theta$) that would be necessary to account for the observed amplitude of ice extent variations $|\eta'|$ in the absence of noise, i.e.,

$$\phi_1 = \frac{|\eta'| - A_\eta \tau_\eta}{A_\theta \tau_\theta \tau_\eta} \qquad [11]$$

If we assume that $A_\theta \approx 3 \times 10^{-4}$ K y^{-1} and that $3 \times 10^2 y < \tau_\theta < 3 \times 10^3 y$, we have from [9],

$$10^{-1} \text{ K} < |\theta'|_e < 1 \text{ K}$$

Because the time constants τ_θ and τ_η are each uncertain to perhaps 1 order of magnitude, it is convenient to rewrite [9] in the logarithmic form

$$\log [\phi_1 \cdot yK] = \log \left[\left(\frac{|\eta'| - A_\eta \tau_\eta}{A_\theta}\right) \cdot y^{-1} K \right] - \log [(\tau_\theta \tau_\eta) \cdot y^{-2}] \quad [12]$$

If we assume $|\eta'| = 0.04$ (~10° lat), $A_\theta = 3 \times 10^{-4}$ Ky^{-1}, $A_\eta = 3 \times 10^{-8}$ y^{-1}, and $\tau_\eta < 10^5 y$, it follows that $A_\eta \tau_\eta \ll |\eta'|$ and

$$\log \left[\left(\frac{|\eta'| - A_\eta \tau_\eta}{A_\theta}\right) \cdot y^{-1} K \right] \approx \log \left[\frac{|\eta'|}{A_\theta} \cdot y^{-1} K \right] \approx 2$$

The relation between ϕ_1 and $(\tau_\theta \tau_\eta)$ is represented by the graph shown in Figure 1. It seems likely that τ_θ and τ_η are each in the neighborhood of $10^3 y$ (i.e., $\log_1[\tau_\theta \tau_\eta \cdot y^{-2}] = 6$). We conclude from this figure that if $\phi_1 = 10^{-4}$ K^{-1} y^{-1} (i.e., $\log [\phi_1 \cdot K y] = -4$) the 100 kyr eccentricity cycle, could, in principle, induce significant variations of marine ice extent, providing these effects are not overwhelmed by omitted nonlinear effects (such as will be discussed in the next section), by forcings of possibly higher amplitude (such as the 20 and 40 kyr orbital variations), or by stochastic noise. It remains to be confirmed from more physical lines of evidence whether values of ϕ_1 in the neighborhood of 10^{-4} K^{-1} y^{-1} are reasonable.

The complete stochastic problem posed by [5] and [6] is more difficult to solve rigorously by analytical methods (cf. (9)) but can easily be solved by numerical methods for sample random forcing. One such solution is shown in Figure 2a,b for the parameters $\tau_\eta = \tau_\theta = 10^3 y$, $A_\eta = 3 \times 10^{-8}$ y^{-1}, $A_\theta = 3 \times 10^{-4}$ K y^{-1}, $|R_\eta| = 10^{-4}$ y^{-1}, and $|R_\theta| = 3 \times 10^{-4}$ K y^{-1}, and $\phi_1 = 10^{-4}$ K^{-1} y^{-1}. The top panel (a) is a sample η'-variation, the dashed curve representing the purely deterministic solution [7]. The middle panel (b) shows the corresponding η'-variation, with the dashed curve again representing the purely deterministic solution [8]. In the last panel (c) we show the η'-variation if ϕ_1 is one order of magnitude lower, 10^{-5} K^{-1} y^{-1}.

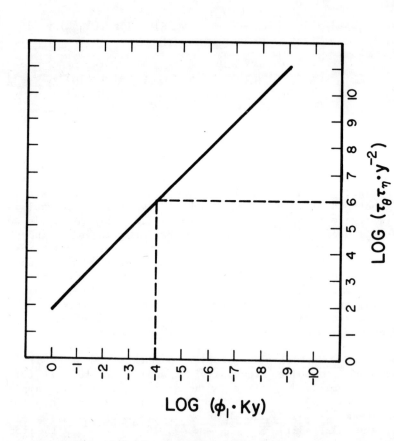

Figure 1 Values of ϕ_1 necessary for eccentricity forcing to induce fluctuations of ice extent of amplitude $|\eta'_1|=0.04$, as a function of the time constant product ($\tau_\theta \tau_\eta$), assuming no stochastic forcing. Values corresponding to $\tau_\eta = \tau_\theta = 10^3$ y are shown by dashed lines.

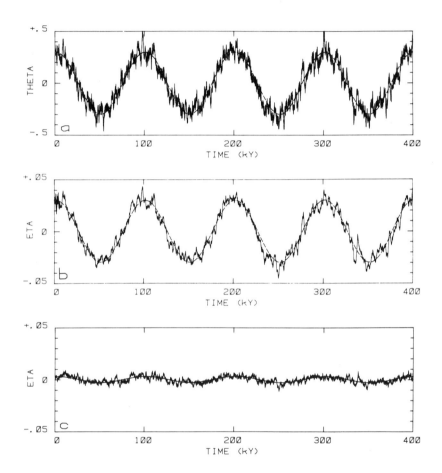

Figure 2 (a) Sample solution for $\theta'(t)$ of the stochastic equation [6] for parameters given in the text. (b) Corresponding solution for $\eta'(t)$ when $\phi_1 = 10^{-4} K^{-1} y^{-1}$, and (c) for $\eta'(t)$ when $\phi_1 = 10^{-5} K^{-1} y^{-1}$.

These numerical solutions are essentially the linear superposition of the deterministic solution given by [7] and [8] (shown by the dashed curve), and the classical Brownian motion stochastic solution of the problem posed by [5] and [6] when the periodic forcing term is omitted (cf., (9)), i.e.,

$$\theta'(t) \approx |\theta'|_e \cos \omega_e t + |\theta'|_R \cdot N(t) \qquad [13]$$

$$\eta'(t) \approx |\eta'|_e \cos \omega_e t + |\eta'|_R \cdot N(t) \qquad [14]$$

where

$$t = n\,\delta t, \quad n = 1, 2, 3, \ldots$$

$$|\theta'|_R = |R_\theta| (2\tau_\theta \delta t)^{1/2} \qquad [15]$$

$$|\eta'|_R = [|R_\eta| + \phi_1 |R_\theta| (2\tau_\theta \delta t)^{1/2}](2\tau_\eta \delta t)^{1/2} \qquad [16]$$

As a final point, we note that if the noise level of the departures of ice extent from the 100 000 year periodic signal can be estimated from observations, it would also be possible to use [16] to obtain an estimate of ϕ_1, i.e.,

$$\phi_1 = \frac{[|\eta|_R - |R_\eta| (2\tau_\eta \delta t)^{1/2}]}{2|R_\theta| (\tau_\theta \tau_\eta)^{1/2} \delta t}$$

CASE II. NONLINEAR, OSCILLATORY FEEDBACK

If we adopt the representations in (1) of both Φ and Ψ, our system takes the form

$$\frac{\delta \eta'}{\delta t} = \phi_1 \theta' - \phi_2 \eta' + A_\eta \cos \omega_e t + R_\eta \qquad [17]$$

$$\frac{\delta \theta'}{\delta t} = -\psi_1 \eta' + \psi_2 \theta' - \psi_3 \eta'^2 \theta' + A_\theta \cos \omega_e t + R_\theta \qquad [18]$$

We shall consider the two sets of values of the coefficients ϕ_1, ϕ_2, ψ_1, ψ_2, and ψ_3 listed in Table 1. In Set A the values of ϕ_1 and ϕ_2 are of the same order as in (1), but ψ_1, ψ_2 and ψ_3 are one order of magnitude smaller. This corresponds to a relaxation time for η of the order of 1 kyr and leads to a limit cycle of somewhat lower amplitude and longer period than in (1) ($\omega_L \gg \omega_e$).

In Set B the values of the coefficients are identical to those specified in (2) (case 2). This corresponds to a long relaxation time for η of the order of 40 kyr, and leads to a limit cycle of period 100 kyr ($\omega_L \approx \omega_e$).

Table 1. Coefficients assigned for the nonlinear dynamical system [17], [18].

coefficient	Values	
	Case A ($\omega \gg \omega_e$)	Case B ($\omega \simeq \omega_e$)
ϕ_1 ($K^{-1}s^{-1}$)	4×10^{-12}	6×10^{-15}
ϕ_2 (s^{-1})	1×10^{-11}	7.5×10^{-13}
ψ_1 (Ks^{-1})	4×10^{-10}	6×10^{-10}
ψ_2 (s^{-1})	2×10^{-11}	1.5×10^{-12}
ψ_3 (s^{-1})	5×10^{-8}	7.5×10^{-8}

Case IIA. ($\omega_L \gg \omega_e$)

In Figure 3a we show the time-domain solutions of [17] and [18] for a sequence of increasing amplitudes, A_θ, ranging from zero to 50×10^{-5} K y^{-1}. As we have noted, plausible values of A_θ lie within these limits. No stochastic forcing is applied in this sequence $|R_\eta| = |R_\theta| = 0$. The values of A_θ and $|R_\theta|$ are given in the lower right hand corner of each panel of these and the following figures in the respective units [10^{-5} K y^{-1}, 10^{-5} K y^{-1}].

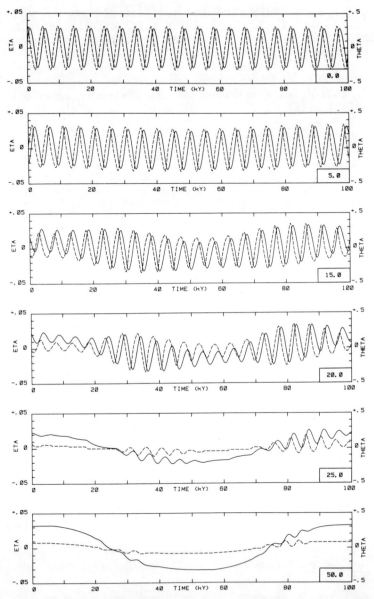

Figure 3 (a) Time domain deterministic solutions for η' (solid cuves) and θ' (dashed curves) of the nonlinear system [17] and [18], for case A (Table 1), for a sequence of increasing amplitudes of eccentricity forcing. Values of A_θ and $|R_\theta|$ are given in the lower right hand corner of each panel in the respective units $[10^{-5} \text{ K y}^{-1}, 10^{-3} \text{ K y}^{-1}]$. In this sequence, $|R_\theta| = 0$.

EARTH-ORBITAL ECCENTRICITY VARIATIONS AND CLIMATIC CHANGE

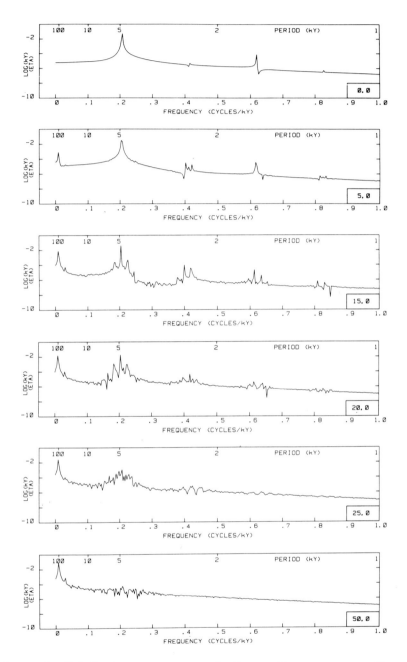

Figure 3 (b) Variance spectra, in log units, corresponding to the η-solutions shown in part (a).

In Figure 3b we show the variance spectra, in log units, corresponding to the time-domain solutions for η given in Figure 3a. The spectra for θ are qualitatively similar and are not shown. In Figure 4a and b we show the set of time and spectral domain results for the same sequence of A_θ-values, but with a noise level $|R_\theta|=3\times10^{-3}$ K y^{-1}.

Finally, in Figures 5a and b we show a set of time and spectral domain results for a fixed value ($A_\theta=50\times10^{-5}$ K y^{-1}) and a variety of additional values of noise amplitude, $|R_\theta|=10^{-3}$, 5×10^{-3} and 10×10^{-3} K y^{-1}.

We note, first, that the purely deterministic, unforced, solution shown in the top panels of Figures 3a and b ($[A_\theta, \epsilon_\theta^{1/2}]=[0,0]$) is a limit cycle of period 4900 years with amplitudes of roughly 0.03 for η (corresponding to an ice edge range of about 7° lat.) and 0.3 K for θ. It is possible to obtain limit cycles of other periods and amplitudes by arbitrarily assigning different values for all the coefficients; we take our particular limit cycle as one example of possible free oscillatory climatic behavior of a relatively short period compared to earth--orbital forcing, in which the variations of η and θ are within plausible limits.

From Figures 3a and b we see that as the amplitude of the 100 000 year forcing is increased from $A_\theta=0$ to $A_\theta=50\times10^{-5}$ K y^{-1} the structure of the solution changes strikingly from one dominated by the free limit cycle to one dominated by the forced long period cycle. For intermediate values of A_θ a subharmonic response occurs, perhaps akin to "period doubling", leading to a destruction of the variance of the free oscillation. The lost variance is not captured by the forcing period, but is redistributed over a broad band of frequencies, reddening the spectrum. From this sequence of solutions we conclude that our free oscillation is extremely sensitive to the amplitude of the long period eccentricity forcing.

However, as shown in other calculations, if the free oscillations of θ were of a much higher amplitude than our above case (e.g., as in (1)) even forcing of amplitude $A_\theta=50\times10^{-5}$ K y^{-1} would cause little change in the limit cycle structure of the solution. Conversely, the signature of a 100 000 year forced oscillation can be effectively obscured by a strong higher frequency internal oscillation.

As noted, the influence of a reasonable level of stochastic noise ($|R_\theta|=3\times10^{-3}$ K y^{-1}) on the solutions shown in Figure 3 is given in Figures 4a and b. The solutions are generally more "realistic" in terms of the general appearance of climatic

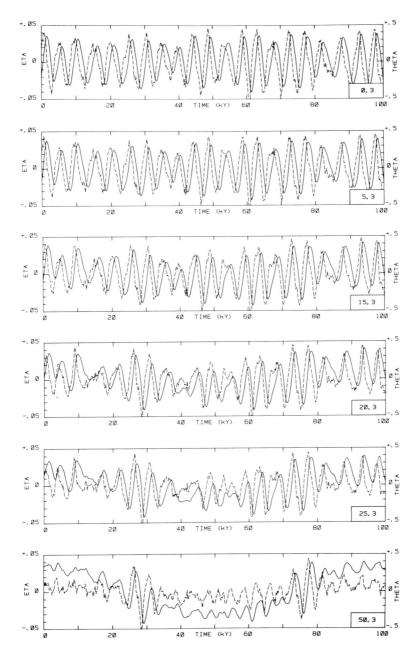

Figure 4a Same as Figure 3a, except that system contains stochastic forcing $|R_\theta| = 3*10^{-3} K\ y^{-1}$.

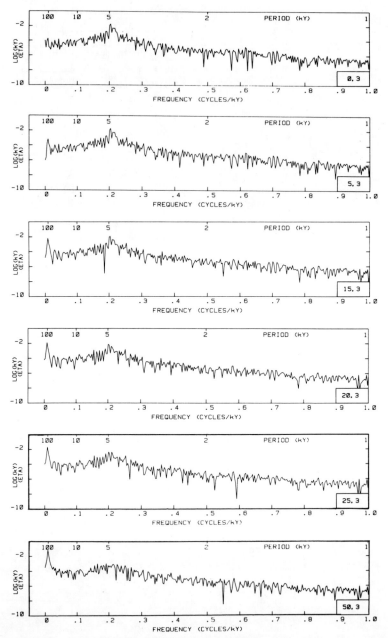

Figure 4b Same as Figure 3 b, except that system contains stochastic forcing $|R_\theta| = 3*10^{-3} K\ y^{-1}$.

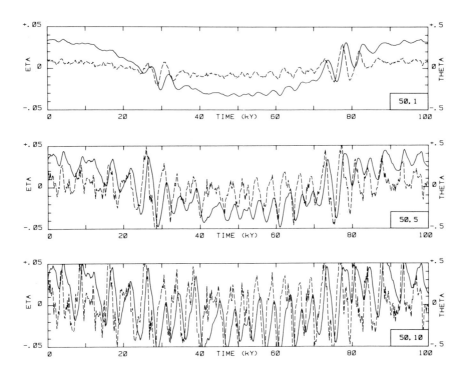

Figure 5a Same as Figure 3a, except that $A_\theta = 50*10^{-5} K\ y^{-1}$ and $|R_\theta| = 10^{-3}$, $5*10^{-3}$, and $10*10^{-3} K\ y^{-1}$ respectively.

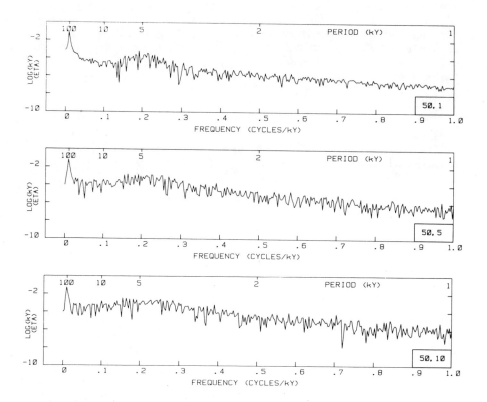

Figure 5b Same as Figure 3 b, except that $A_\theta = 50*10^{-5} K\ y^{-1}$ and $|R_\theta| = 10^{-3}$, $5*10^{-3}$, and $10*10^{-3} K\ y^{-1}$ respectively.

records. In all cases the noise level is not too large to obscure the main deterministic flow of the trajectory. In the high amplitude forcing case of $A_\theta = 50 \times 10^{-5}$ K y^{-1}, we can observe that the noise has the interesting effect of interfering with the forcing enough to permit a stronger limit cycle signal than was possible in purely deterministic case (cf. Figure 3), i.e., the noise increases the structural stability of the limit cycle. The effects of even higher amplitudes of noise $|R_\theta| = 5 \times 10^{-3}$ and 10×10^{-3} K y^{-1} are shown in the last two panels of Figures 4a and b, respectively.

Case IIB ($\omega_L \simeq \omega_e$)

As noted by J. Imbrie in this symposium, an analysis of the phase spectrum between the orbital forcing and the oxygen isotope/ice volume signal reveals a strong possibility of resonance at the eccentricity period of 100 kyr. It is implied that there must be some free behavior at this period in the system. As a prototype of this possibility, let us examine the response to 100 kyr eccentricity forcing of the 100 kyr limit cycle solution obtained from our model in (2). The coefficients are listed in Table 1 (Case B). The results are shown in Figure 6, the top panel of which shows the previous, unforced, deterministic results ($A_\theta = |R_\theta| = 0$).

In the next two panels we show the results when we add 100 kyr periodic forcing with $A_\theta = 8 \times 10^{-5}$ K y^{-1} (the dotted line shows the phase of the periodic forcing, $A_\theta \cos \omega_e t$, with arbitrary amplitude), and with this forcing plus stochastic noise $|R_\theta| = 3 \times 10^{-3}$ K y^{-1} as before, respectively. We note some amplification of the θ-oscillations, and a fixing of the phase of ice extent relative to the periodic forcing that is qualitatively similar to the observations as portrayed in (10, Figure 10).

CONCLUSIONS

Our calculations, based on the marine ice/deep ocean temperature model, lead to the following main results:

(1) For a large enough sensitivity, ϕ_1, of ice mass change to bulk ocean temperature it is possible that there can be a direct forcing of the ice mass by the 100 kyr eccentricity variations. The response can then be consistent with observations (e.g., (10)) and with simple energy considerations (11,12) which show that the latent heats of fusion associated with the known variations of ice sheets are of the same order as the amplitude of the eccentricity forcing.

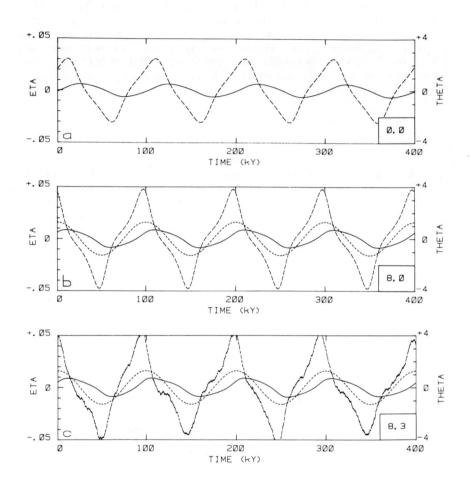

Figure 6 :
  <u>top</u> Time domain deterministic solution of the nonlinear system [17] and [18] for case B (Table 1), with no forcing. Solid curve is η' and dashed line is θ'. From [12].
  <u>middle</u> Deterministic solutions, with periodic forcing of amplitude $A_\theta = 8*10^{-5}$ (shown, with arbitrary amplitude, as the dotted curve).
  <u>bottom</u> Solution with periodic forcing as above, plus stochastic noise of amplitude $|R_\theta| = 3*10^{-3} K\ y^{-1}$.

(2) As would be expected, this forced response can be obscured if free oscillations of shorter period have a high amplitude.

(3) Moreover, the forced eccentricity response is not structurally stable even in the presence of a relatively weak short period oscillation of the kind shown in Figure 3 (top panel).

(4) From the converse view point, short period free oscillations characterized by θ-variations of the order of 10^{-1} K are extremely sensitive to small changes of amplitude of the 100 kyr eccentricity forcing that are well within the limit of error of this eccentricity amplitude, i.e., within the limits of uncertainty of the eccentricity forcing there can be a wide variety of responses in the presence of a reasonable level of shorter period auto-oscillation. See Figure 3.

(5) If a free oscillation of very long period (e.g., 100 kyr) is admitted by the climatic system, this oscillation can be amplified somewhat and its phase resonantly fixed by periodic forcing of the same period. The resulting signal is not unrealistic when compared to the observed ice volume record (compare (10)).

(6) Aside from leading to responses that appear more realistic, the presence of a reasonable level of stochastic noise renders limit cycle behavior of shorter period more stable to periodic 100 kyr forcing than would otherwise be the case. Compare Figures 3a and 4a.

ACKNOWLEDGEMENTS

This material is based upon work supported by the Division of Atmospheric Sciences, National Science Foundation, under Grant ATM-7925013 at Yale University, and Grants ATM-8022869 and ATM-8213094 at the Center for the Environment and Man, Inc.

REFERENCES

1. Saltzman, B., Sutera, A., and Evenson, A. : 1981, J. Atmos. Sci. 38, pp. 494-503.
2. Saltzman, B., Sutera, A., and Hansen, A.R. : 1982, J. Atmos. Sci. 39, pp. 2634-2637.
3. Milankovitch, M. : 1930, Handbuch der Klimatologie, Berlin, Gebrüder Bornträger, 1, A, 176pp.
4. Berger, A.L. : 1977, Nature 269, pp. 44-45
5. Saltzman, B. : 1978, Advances in Geophysics 20, pp. 183-304.

6. Le Treut, H., and Ghil, M. : 1983, J. Geophys. Res. 88, pp. 5167-5190.
7. North, G.R., Cahalan, R.F., and Coakley, J.A.Jr. : 1981, Rev. Geophys. and Space Phys. 19, pp. 91-121.
8. Saltzman, B., and Moritz, R.E. : 1980, Tellus 32, pp. 99-118.
9. Chandrasekhar, S. : 1943, Rev. Mod. Phys. 15, pp. 1-91.
10. Imbrie, J., and Imbrie, J.Z. : 1980, Science 207, pp. 943-953.
11. Fong, P. : 1982, Climatic Change 4, pp. 199-206.
12. Saltzman, B. : 1977, Tellus 29, pp. 205-212.

SELF-OSCILLATIONS, EXTERNAL FORCINGS, AND CLIMATE PREDICTABILITY

C. Nicolis

Institut d'Aéronomie Spatiale de Belgique, B-1180
Brussels, Belgium

ABSTRACT

A class of nonlinear climate models involving two simultaneously stable states, one stationary and one time-periodic, is analyzed. The evolution is cast in a universal, "normal form", from which a basic difference between "radial" and "phase" variables emerges. In particular, it is shown that the phase variable has poor stability properties which are at the origin of progressive loss of predictability when the oscillator is autonomous. Next, the coupling to an external periodic forcing is considered. New types of solution are found, which correspond to a sharp and reproducible behavior of the phase. In this way predictability is ensured. Moreover, it is shown that the response can be considerably amplified by a mechanism of resonance with certain harmonics of the forcing. The implication of the results on the mechanism of quaternary glaciations is emphasized.

INTRODUCTION

A central question in climate dynamics concerns the relative role of internally generated and external mechanisms of climatic change. It is well known that many climatic episodes, among which quaternary glaciations are the most striking example, present a cyclic character whose characteristic time is strongly correlated with external periodicities, like those of the earth's orbital variations (1). On the other hand, the balance equations of the principal variables predict that the coupling of such external forcings to the system's dynamics is exceedingly small. Hence, it is difficult to understand how such small amplitude disturbances can cause a response in the form of a major climatic change.

In the last few years two types of explanation of this "apparent paradox" have been advanced. In a first attempt by Benzi et al. (2) and the present author (3), the coupling between an energy balance model giving rise to multiple stable steady states and an orbital forcing perturbing periodically the solar constant has been considered. It has been shown that the <u>internal fluctuations</u> generated spontaneously by the system can amplify dramatically the response to the forcing, provided that the period of the latter is comparable to the mean first passage time between the stable climatic states. As a result, the system experiences systematic deviations from the present-day climate to a less favorable climate, which are entrained to the periodicity of the external forcing.

A second attempt at a solution of the problem is to examine to what extent the climatic system is capable of generating <u>self-oscillating</u> dynamical behavior. Saltzman and coworkers (4,5,6,7) analyzed the interactions between sea ice extent and ocean surface temperature, and revealed the existence of an autonomous oscillator arising from the insulating effect of sea ice on temperature, and from the negative effect of temperature on sea ice. For plausible parameter values the periodicity turns out to be of the order of 10^3 years. This was to be expected, in view of the fact that such a value is an upper bound of variation in time of sea ice extent (8). On the other hand, by analyzing the interactions between the meridional ice-sheet extent and global temperature, Ghil and co-authors (9,10,11,12) have also identified an autonomous oscillator. The periodicity of the latter turns out to be larger compared to Saltzman's oscillator, of the order of 10^4 years for realistic parameter values.

In a recent paper devoted to the implications of self-oscillations in climate dynamics (13), we arrived at the surprising conclusion that an autonomous oscillator subject to its own internal fluctuations or to a noisy environment has poor predictability properties. Specifically, its dynamics can be decomposed into a "radial" part and an "angular" part. It then turns out that the radial variable has strong stability properties, whereas the phase variable becomes completely deregulated by the fluctuations. As a result, any trace of the oscillatory behavior is wiped out after a sufficiently long lapse of time. It would appear therefore that self-oscillations alone cannot provide a reliable mechanism of cyclic climatic change encompassing both a long time interval and a global space scale.

The purpose of the present work is to show that the coupling between an autonomous climatic oscillator and an external periodic forcing can, under certain conditions, ensure sharp predictability. The latter will be reflected by the occurrence of cyclic variations of climate displaying a well-defined periodicity. We focus on the role of orbital forcings in quaternary glaciations. As the relevant orbital periodicities are of the order of 2×10^4 to 10^5 years, it is clear that only land-ice dynamics is expected to couple effectively to such variations. For this reason we subsequently consider the type of models analyzed by Ghil (12) as a prototype of this coupling. In the next section, we cast these models in a generic, "normal" form and summarize the results concerning the bifurcation of various kinds of solution. We next show how poor predictability can arise in the absence of external forcing. The remaining part of the work is devoted to the effect of such a forcing. We show that, contrary to the autonomous oscillator, the driven one attains, asymptotically, a well defined phase. Moreover, substantial amplification of the response can take place by the occurrence of resonance with some of the harmonics of the forcing. Finally, in the last section, we present some comments on the implications of the results.

NORMAL FORM AND BIFURCATION ANALYSIS OF GHIL TYPE MODELS

In Ghil and Tavantzis (12) a set of two coupled nonlinear differential equations for the evolution in time of the dimensionless variables θ, ℓ descriptive of the mean surface temperature and the meridional ice-sheet extent is analyzed. For plausible values of the physical parameters a stable steady-state climate is shown to coexist with a stable time-periodic one. An intermediate unstable periodic branch separates the above two solutions. The amplitude of the stable periodic solution, which emerges by a mechanism of Hopf bifurcation, corresponds roughly to that obtained from data on quaternary glaciation cycles, whereas its period is in the 10^4 year range. For other parameter values, more complex behavior becomes possible, culminating in the appearance of infinite period orbits represented in phase space by a separatrix loop which is doubly asymptotic to an unstable steady state. We are not interested here in this latter type of behavior. Instead, we focus on the range of two stable solutions, one of them periodic, separated by an unstable solution.

According to the qualitative theory of ordinary differential equations (14) any system operating in the vicinity of a Hopf bifurcation and presenting the above mentioned bistable behavior can be transformed, by a suitable change of variables, to the following universal form, known as normal form:

$$\frac{\partial z}{\partial t} = z\,(i\omega_0 + \beta + \gamma|z|^2 - \zeta|z|^4) \quad [1]$$

Here z, γ, ζ are complex-valued combinations of the initial variables and parameters; β measures the distance from the Hopf bifurcation; and ω_0 is the frequency of the periodic solution at exactly the bifurcation point ($\beta=0$). For simplicity we hereafter take γ and ζ to be real, and set without further loss of generality $\zeta=1$. Instead of viewing eq.[1] as equivalent to a pair of equations for the real and imaginary parts of z (which, as already pointed out, are suitable combinations of the initial balance equations), it now becomes convenient to switch to "radial" and "angular" variables through

$$z = r\,e^{i\phi} \quad [2]$$

Substituting into eq.[1] and equating real and imaginary parts on both sides we obtain :

$$\frac{dr}{dt} = r\,(\beta + \gamma r^2 - r^4) \quad [3a]$$

$$\frac{d\phi}{dt} = \omega_0 \quad [3b]$$

This set of equations admits a single steady state solution,

$$r_{so} = 0 \quad [4]$$

It also admits solutions in which r is a nontrivial root of the right hand side of [3a] and ϕ varies according to [3b] :

$$\phi = \omega_0 t + \phi_0 \quad [5]$$

ϕ_0 being an (arbitrary) initial phase. In view of eq.[2], these are therefore time periodic solutions of the initial problem. To determine them we consider the biquadratic

$$r_s^4 - \gamma r_s^2 - \beta = 0$$

or

$$r_{s\pm}^2 = \frac{1}{2}[\gamma \pm (\gamma^2 + 4\beta)^{1/2}] \quad [6]$$

This admits two positive solutions as long as the following inequalities are satisfied :

$$-\frac{\gamma^2}{4} \leqslant \beta \leqslant 0\,,\ \gamma > 0 \quad [7]$$

Linear stability analysis carried out on eq.[3a] shows that r_{so} is stable for $\beta<0$ and unstable for $\beta>0$, r_{s+} is stable whenever it exists ($\beta \geqslant -\gamma^2/4$), and r_{s-} is always unstable. Figures 1a,b summarize the information concerning these various solutions in the form of a <u>bifurcation diagram</u>, completed by a <u>state diagram</u> in parameter space.

The resemblance between Figure 1a and the results of Ghil and Tavantzis (12) is striking. Actually we can make use of the formal identity of the results to fit some of the parameter values of our model equation. Specifically, in the Ghil-Tavantzis analysis bifurcation occurs when μ, the ratio of the heat capacity to the land ice inertia coefficient, is equal to $\mu = 1.76735$. Moreover the two periodic solutions -the analogs of our r_{s+} and r_{s-} disappear below $\mu = 1.7583$. Comparing with inequalities [7] we see that our equation can fit this model provided that

$$\beta = \mu - 1.76735$$

and

$$\frac{\gamma^2}{4} = 1.76735 - 1.7583$$

or

$$\gamma = 0.1903 \qquad [8a]$$

Note that the present day climate in Ghil-Tavantzis analysis corresponds to $\mu = 1.76$ or, in our notation, to

$$\beta_s = -0.00735 \qquad [8b]$$

From relations [8] we arrive at the rather interesting conclusion that, for realistic parameter values, we can consider that our model operates close to the bifurcation point $\beta = 0$. Moreover, the width of the multiple stable state region as measured by γ is small. As we see in the next sections these features will allow us to perform a systematic analytical study of the system.

STOCHASTIC PERTURBATIONS AND PREDICTABILITY

The decomposition of the dynamics into a radial and phase part achieved by eqs.[3a] - [3b] allows us to obtain, straightforwardly, preliminary information on the predictability properties of our model oscillator. Indeed, consider the following thought experiment. Suppose that the system runs on its limit cycle $r = r_{s+}$. At some moment, corresponding to a value $\phi = \phi_1$ of the phase, we displace the system to a new state characterized by the values r_0, ϕ_0 of the variables r and ϕ. According to

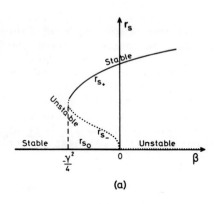

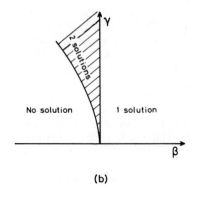

Figure 1a Bifurcation diagram for the amplitude of eq. [1] as a function of the parameter β. Full and dotted lines denote respectively stable and unstable branches.

Figure 1b Dashed region indicates the domain of parameter space β, γ for which there are two non-trivial periodic solutions of eq. [1].

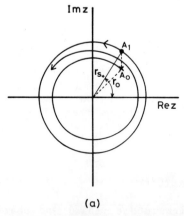

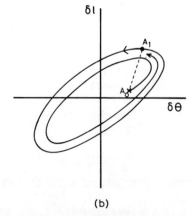

Figure 2 Schematic representations of the evolution following the action of a perturbation leading from state A_1 on the limit cycle to state A_0. Parts (a) and (b) describe the situation, respectively, in the space of the variables of normal form and in the space of the variables, δl, $\delta \theta$ representing the deviations of the original variables l, θ from the steady state.

eq.[3a] and the stability analysis performed in the previous section the variable r will relax from r_0 back to the value r_{s+}, as the representative point in phase space will spiral toward the limit cycle (cf. Fig. 2). On the other hand, according to eq.[5], the phase variable ϕ will keep forever the memory of the initial value ϕ_0. In other words, when the limit cycle will be reached again, the phase will generally be different from the one that would characterise an unperturbed system following its limit cycle during the same time interval. In as much as the state into which the system can be thrown by a perturbation is unpredictable, it therefore follows that the reset phase of the oscillator will also be unpredictable. In other words, <u>our non-linear oscillator is bound to behave sooner or later in an erratic way</u> under the action of perturbations. This is tantamount to poor predictability.

The above surprising property can be further substantiated by a stochastic analysis of the model. As well known any complex physical system possesses a universal mechanism of perturbations generated spontaneously by the dynamics, namely the fluctuations. Basically, fluctuations are random events. It thus follows that the state variables themselves (r,ϕ) (or θ and ℓ in Ghil's original model) become random processes. As we will show shortly this will result in a complete deregulation of the phase variable, whereas the radial variable will remain robust. The end result will be that because of destructive phase interference (cf. eqs.[2] and [5] with a randomly distributed initial phase ϕ_0), the signal of the system will tend to become flat if averaged over a sufficiently long time.

The analysis of fluctuations follows the same lines as in Nicolis (13). We incorporate the effect of fluctuations by adding random forces F_θ, F_ℓ to the deterministic rate equations (15). As usual, we assume the latter to define a multi-Gaussian white noise :

$$< F_\theta(t) \, F_\theta(t') > = q_\theta^2 \, \partial(t - t')$$

$$< F_\ell(t) \, F_\ell(t') > = q_\ell^2 \, \partial(t - t')$$

$$< F_\theta(t) \, F_\theta(t') > = q_{\theta\ell} \, \partial(t - t') \quad\quad [9]$$

This allows us to write a Fokker-Planck equation for the probability disitribution $P(\theta,\ell,t)$ of the climatic variables. In general this equation is intractable. However, the situation is greatly simplified if one limits the analysis to the range in which the normal form (eqs.[1] and [3]) is valid. To see this we first express the Fokker-Planck equation in the polar coordinates (r,ϕ). We obtain [13,16]

$$\frac{\partial P(r,\phi,t)}{\partial t} = -\frac{\partial}{\partial r}\{(\beta + \gamma r^2 - r^4)r + \frac{1}{2r}Q_{\phi\phi}\}P$$

$$-\frac{\partial}{\partial \phi}\{\omega_0 - \frac{1}{r^2}Q_{r\phi}\}P$$

$$+\frac{1}{2}\{\frac{\partial^2}{\partial r^2}Q_{rr} + 2\frac{\partial^2}{\partial r \partial \phi}\frac{Q_{r\phi}}{r} + \frac{\partial^2}{\partial \theta^2}\frac{Q_{\phi\phi}}{r^2}\}P \quad [10]$$

in which $Q_{\phi\phi}$, $Q_{r\phi}$, Q_{rr} are suitable combinations of q_ϕ^2, q_ℓ^2, $q_{\theta\ell}$, $\sin\phi$ and $\cos\phi$.

To go further it is necessary to introduce a perturbation parameter in the problem. We choose it to be related to the weakness of the noise terms, and we express this through the scaling

$$Q_{\phi\phi} = \varepsilon \tilde{Q}_{\phi\phi}$$

$$Q_{r\phi} = \varepsilon \tilde{Q}_{r\phi}$$

$$Q_{rr} = \varepsilon \tilde{Q}_{rr}, \quad \varepsilon \ll 1 \quad [11]$$

We next recall that, in view of the remark made earlier, both the bifurcation parameter β and the parameter γ controlling the width of the multiple stable state region can be taken small. We express this by scaling these parameters by suitable powers of ε, chosen in such a way that the scaled Fokker-Planck equation still admits smooth and non-trivial solutions. After a long calculation (13,17) we obtain the following two results.

(i) Let us define the conditional distribution $P(\phi/r,t)$ through

$$P(r,\phi,t) = P(\phi/r,t) P(r,t) \quad [12]$$

Then, to dominant order in ε, $P(\phi/r,t)$ obeys to the equation

$$\frac{\partial P(\phi/r,t)}{\partial t} = -\frac{\partial}{\partial \phi}\omega_0 (P\phi/r,t) \quad [13a]$$

which admits a properly normalized stationary solution

$$P_s(\phi/r) = \frac{1}{2\pi} \qquad [13b]$$

(ii) Substituting [13b] into the bivariate Fokker-Planck equation for $P(r,\phi,t)$ and keeping dominant terms in ϵ we find a closed equation for $P(r,t)$ of the form :

$$\frac{\partial P(r,t)}{\partial t} = -\frac{\partial}{\partial r}(\beta r + \gamma r^3 - r^5 + \frac{Q}{2r}) + \frac{Q}{2}\frac{\partial^2}{\partial r^2}P \qquad [14]$$

in which the positive definite quantity Q is another combination of q_θ^2, q_ℓ^2, $q_{\theta\ell}$. Eq.[14] admits the following steady-state solution :

$$P_s(r) \sim r \exp[-\frac{2}{Q}U(r;\beta,\gamma)] \qquad [15a]$$

where $U(r;\beta,\gamma)$ is the <u>kinetic potential</u> generating the equation of evolution for r :

$$\frac{dr}{dt} = -\frac{\partial U}{\partial r}$$

From eq.[3a] :
$$U(r;\beta,\gamma) = -\beta\frac{r^2}{2} - \gamma\frac{r^4}{4} + \frac{r^6}{6} \qquad [15b]$$

Figure 3 represents the function [15a] in the range $-\gamma^2/4 < \beta < 0$. We obtain a distribution in the form of a wine bottle whose upper part has been cut. The projection of the upper edge and of the basis of the bottle on the phase plane are, respectively, the stable and unstable deterministic limit cycles.

The main conclusion to be drawn from the above analysis is that the phase variable has a completely flat probability distribution (cf. eq.[13b]). It may therefore be qualified as "chaotic", in the sense that the dispersion around its average will be of the same order as the average value itself. On the other hand, the radial variable has a stationary distribution (cf. eq.[15a]) such that the dispersion around the most probable value ($r=0$ or $r=r_{s+}$) is small. Nevertheless, the mere fact that the probability distribution is stationary rather than time-periodic implies that a remnant of the chaotic behavior of ϕ subsists in the statistics of r : if an average over a large number of samples (or over a sufficient time interval in a single realization of the stochastic process) is taken, the perio-

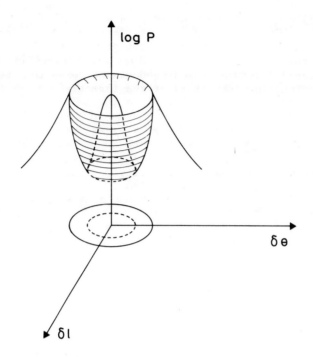

Figure 3 schematic representation of the steady state probability distribution eqs [15a] - [15b] as a function of the excess variables δl and $\delta \theta$.

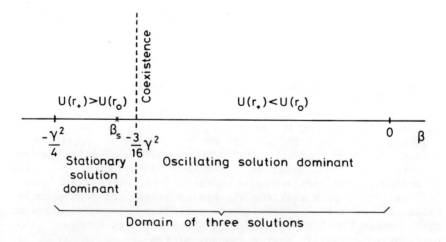

Figure 4 Range of values of the parameter β for which the stationary solution or the time periodic solution is the most probable state of the system.

dicity predicted by the deterministic analysis will be wiped out as a result of destructive phase interference. This property is at the origin of a progressive loss of predictability.

Using eqs.[15] one can also determine the range of parameter values over which the steady state climate (r=0) or the time periodic one (r=r_{s+}) are dominant, in the sense that they correspond to the deepest of the two minima of the potential. The results are summarized in Figure 4. As expected, near the bifurcation point $\beta=0$ the time-periodic solution dominates. On the contrary, near the limit point $\beta=-\gamma^2/4$ the steady-state climate is the most probable one.

The knowledge of the steady-state disitribution [15a] is also sufficient for calculating the mean characteristic passage times needed to jump between the two stable states through fluctuations (18). We find (17) that for the parameter values fitting present climate in Ghil's model (see eqs [8]), this time is only twice the periodicity of the limit cycle itself. This is another manifestation of the poor predictability properties of the system.

EFFECT OF A PERIODIC FORCING : PREDICTABILITY ESTABLISHED

We now consider the response of the climatic oscillator described by eqs.[1] and [3] to an external periodic forcing. Our purpose is to show that the forcing provides the synchronizing element that was missing in the autonomous evolution, and ensures in this way the existence of a sharp and predictable signal.

We analyze here for simplicity a purely sinusoidal forcing. The most general coupling with the internal dynamics would be described by an augmented eq.[1], in which both additive contributions as well as contributions multiplied by suitable powers of z are considered. It is clear however that, as long as the system is in the range of small β and γ, $|z|$ itself would be small. As a result the response will be dominated by the additive part.

To study quantitatively the effect of the forcing it will be convenient to work with the equations for the real and imaginary parts of our variable z, in view of the subtleties associated with the handling of the phase variable. We can write these equations in the form (cf. eq.[1] with z=x+iy) :

$$\frac{dx}{dt} = \beta x - \omega_0 y + \gamma x(x^2 + y^2) - x(x^2 + y^2)^2 + \delta q \sin \omega_1 t$$

$$\frac{dy}{dt} = \omega_0 x + \beta y + \gamma y (x^2 + y^2) - y(x^2 + y^2)^2 \qquad [16]$$

Here ω_1 is the frequency of the foring, δ a smallness parameter, q the coupling amplitude. The absence of coupling in the second eq.[16] simplifies the calculations considerably and will therefore be adopted in the sequel. Note that there is no essential loss of generality implied by such an assumption.

In order to handle eqs.[16], we shall take into account the fact, already utilized previously, that the system operates close to bifurcation ($\beta \to 0, \gamma \to 0$). Setting

$$x = \delta x_1 + \delta^2 x_2 + \delta^3 x_3 + \dots$$

$$y = \delta y_1 + \delta^2 y_2 + \delta^3 y_3 + \dots \qquad [17]$$

we then see that to order δ the only terms surviving in [16] are

$$\frac{dx_1}{dt} + \omega_0 y_1 = q \sin \omega_1 t$$

$$\frac{dy_1}{dt} - \omega_0 x_1 = 0 \qquad [18]$$

A particular solution of [18] is easily found to be

$$x_{1P} = \frac{q\omega_1}{\omega_0^2 - \omega_1^2} \cos \omega_1 t$$

$$y_{1P} = \frac{q\omega_0}{\omega_0^2 - \omega_1^2} \sin \omega_1 t \qquad [19]$$

This solution is well-behaved for all values of ω_1, except those for which there is resonance, $\omega_1 \sim \omega_0$. In the context of quaternary glaciations, in which the frequency of the (orbital) forcing is at least two times smaller than ω_0, it is unlikely that resonance can be expected. We therefore exclude this possibility for the time being.

The general solution of eqs.[18] is given by the particular solution eq.[19] to which the general solution of the homogeneous equation is added. As the homogeneous equation is simply the harmonic oscillator problem, we finally obtain :

$$x_1 = A \cos \omega_0 t + B \sin \omega_0 t + \frac{q \omega_1}{\omega_0^2 - \omega_1^2} \cos \omega_1 t$$

$$y_1 = A \sin \omega_0 t - B \cos \omega_0 t + \frac{q \omega_0}{\omega_0^2 - \omega_1^2} \sin \omega_1 t \qquad [20]$$

At this point A and B are undetermined constants. As a matter of fact their indeterminacy reflects, in part, the lability of the phase variable in the absence of the forcing. Contrary to the autonomous case however, we now have a way to remove this indeterminacy. It suffices to push expansion [17] to a higher order with the additional requirement that β and γ have also to be scaled in terms of δ :

$$\beta = \delta^4 \tilde{\beta} + \ldots$$

$$\gamma = \delta^2 \tilde{\gamma} + \ldots \qquad [21]$$

We obtain in this way, to the first nontrivial order beyond (x_1, y_1) :

$$\frac{dx_5}{dt} + \omega_0 y_5 = \tilde{\beta} x_1 + \tilde{\gamma} x_1 (x_1^2 + y_1^2) - x_1 (x_1^2 + y_1^2)^2$$

$$\frac{dy_5}{dt} - \omega_0 x_5 = \tilde{\beta} y_1 + \tilde{\gamma} y_1 (x_1^2 + y_1^2) - y_1 (x_1^2 + y_1^2)^2 \qquad [22]$$

in which x_1, y_1 are given by eqs.[20].

This inhomogeneous set of equations for (x_5, y_5) admits a solution only if a <u>solvability condition</u> expressing the absence of secular terms (i.e. terms increasing unboundedly in t) is satisfied. It turns out (19) that this condition expresses the orthogonality of the right hand side of [22], viewed as a vector, to the two eigenvectors of the operator in the left hand side :

$$(\cos \omega_0 t, \sin \omega_0 t)$$

and

$$(\sin \omega_0 t, - \cos \omega_0 t) \qquad [23]$$

The scalar product to be used is the conventional scalar product of vector analysis, supplemented by an averaging over t. The point is that we dispose of two such solvability conditions, for the two unknown constants A and B. Because of the presence of terms in q in eq.[20] these will be inhomogeneous equations. Both A and B will therefore be fixed entirely, without further indeterminacy. In other words, whatever the initial perturbation which may act on the system, the final solution will be perfectly well defined both as far as its amplitude and its phase are concerned. We have therefore shown that the unpredictability pointed out in the previous section is removed by the presence of the forcing.

Once the solvability condition is satisfied one can compute (x_5, y_5) from eq.[22]. A novel feature then appears since the solution involves trigonometric functions having arguments $3\omega_1 t$ and $5\omega_1 t$. This will give rise to denominators of the form $(\omega_0 - 3\omega_1)$ and $(\omega_0 - 5\omega_1)$. The appearance of resonance becomes now much more plausible, since for an intrinsic period of say 7×10^3 years it would be realized by external periodicities of 2×10^4 and 4×10^4 years. These are known to be present in the orbital forcing. In other words, in addition to establishing predictability we obtain, the enhancement of the response in the form of resonance with certain harmonics of the external forcing.

DISCUSSION

The principal result of the present paper has been that self-oscillations in climate dynamics are bound to show erratic behavior after the lapse of sufficiently long time, as a result of poor stability properties of the phase variable. On the other hand when the nonlinear oscillator is coupled to an external periodic forcing the response is characterized, under typical conditions, by a sharply defined amplitude and phase. In other words the signal becomes "predictable" in the sense that its power spectrum computed from a time series of data would be dominated by a limited number of well-defined frequencies.

In addition to predictability the coupling with the external forcing can also lead to enhancement of the response, through a mechanism of resonance between the intrinsic frequency and certain harmonics of the forcing. Besides, eqs [20] show that the response to the forcing will in general be quasi-periodic, in view of the fact that the internal and external frequencies need not be rationally related to each other. These features are in complete agreement with the numerical simulations by Le Treut and Ghil (11).

In the analysis of the effect of a periodic forcing, we adopted a purely phenomenological description. A stochastic analysis incorporating the effect of both fluctuations and external forcing, would certainly provide a more convincing proof of how predictability is ensured in the system. We intend to report on this point in future investigations.

ACKNOWLEDGEMENTS

I wish to thank M. Ghil for a critical reading of the manuscript. This work is supported, in part, by the EEC under contract n° CLI-027-B(G).

REFERENCES

1. Berger, A. : 1978, Quaternary Research 9, pp. 139-167.
2. Benzi, R., Parisi, G., Sutera, A., and Vulpiani, A. : 1982, Tellus 34, pp. 10-16.
3. Nicolis, C. : 1982, Tellus 34, pp. 1-9.
4. Saltzman, B. : 1978, Adv. Geophys. 20, pp. 183-304.
5. Saltzman, B., and Moritz, R.E. : 1980, Tellus 32, pp. 93-118.
6. Saltzman, B., Sutera, A., and Evenson, A. : 1981, J. Atmos. Sci. 38, pp. 494-503.
7. Saltzman, B. : 1982, Tellus 34, pp. 97-112.
8. Berger, A. : 1981, in : "Climatic Variations and Variability Facts and Theories", A. Berger (Ed.), Reidel Publ. Company, Holland, pp. 411-432.
9. Källén, E., Crafoord, C., and Ghil, M. : 1979, J. Atmos. Sci. 36, pp. 2292-2303.
10. Ghil, M., and Le Treut, H. : 1981, J. Geophys. Res. 86C, pp. 5262-5270.
11. Le Treut, H., and Ghil, M. : 1983, J. Geophys. Res., in press.
12. Ghil, M., and Tavantzis, J. : 1983, SIAM J. Appl. Math., in press.
13. Nicolis, C. : 1983, Tellus, in press.
14. Arnold, V. : 1980, Chapitres supplémentaires de la théorie des équations différentielles ordinaires. Mir, Moscow.
15. Hasselmann, K. : 1976, Tellus 28, pp. 473-485.
16. Baras, F., Malek-Mansour, M., and Van den Broeck, C. : 1982, J. Stat. Phys. 28, pp. 577-587.
17. Nicolis, C. : 1983, to be published.
18. Nicolis, C., and Nicolis, G. : 1981, Tellus 33, pp. 225-237.

19. Sattinger, D. : 1973, Topics in Stability and Bifurcation Theory. Springer-Verlag, Berlin.

SENSITIVITY OF INTERNALLY-GENERATED CLIMATE OSCILLATIONS TO OCEAN MODEL FORMULATION

L.D.D. Harvey[1] and S.H. Schneider

National Center for Atmospheric Research[2], Boulder, CO 80307, USA.

ABSTRACT

   The simulation of climate on Pleistocene time scales requires coupling subsystems with response time scales which vary over several orders of magnitude. Most models of the coupled atmosphere-ocean-cryosphere (AOC) system have treated the global ocean as a single isothermal reservoir, resulting in a relatively slow surface temperature response time scale. We use a series of one-dimensional box ocean models and a box-advection diffusion ocean model to demonstrate that the ocean temperature does not respond with a single time scale. In particular, the first half or more of the mixed layer response to a radiative perturbation is governed by the relatively small mixed layer thermal inertia, rather than by the deep ocean thermal inertia. We implement the box and box-advection-diffusion ocean models in the AOC model of (7), which uses a single isothermal ocean reservoir and exhibits internal oscillations. We find that internal oscillations are no longer a solution to their model, thereby confirming the importance of distinguishing between the mixed layer and deeper ocean thermal response time scales.

INTRODUCTION

   Simulating climate on Pleistocene time scales from a climate modelling point of view requires coupling of subsystems with characteristic thermal response time scales which vary over several orders of magnitude. Solutions to models of coupled subsystems range from one steady state solution for each external forcing to internal oscillations occurring among the climate subsystems even with constant external forcing. The nature of the solution depends, of course, on both the functional form of the physical parameterizations and the numerical values of the physical coefficients which determine the magnitude of the energy and/or mass flows between different components of the system or to space. Processes occurring on time scales much less than the Milankovitch time scale can nonetheless have a profound impact on the character of solutions to coupled climate system models on Pleistocene time scales. Internal oscillations, which are solutions to some models, can be eliminated by plausible variations in both the physical formulation and in the parameter values. For example, we show here that different assumptions for ocean mixing processes on $10^1 - 10^2$ year time scales can eliminate oscillations in coupled atmosphere-ocean--cryosphere (AOC) models.

Theories proposed to explain the ice ages of recent geological history can be divided into those which invoke external forcing as the driving mechanism, those which invoke free internal oscillations without any external forcing, and those which invoke a resonance between internal oscillation frequencies and external forcing frequencies. An important feature of the latter two categories of theories is that the components of the climate system are in general in a state of disequilibrium which creates continuously changing energy flows among them. Also, in the model of Kallen, Crawford and Ghil (7) this disequilibrium occurs at a global scale in the radiative balance term as the difference between the absorbed solar and outgoing infrared radiation terms oscillates with an amplitude of 6 W m^{-2} (equivalent to the change in the annual absorbed radiation associated with a 3% solar constant variation). Among the external forcing theories, on the other hand, consider the example of Pollard (8). He integrates a glacier model continuously for periods of 2000 years with the atmosphere seasonal cycle constant, then computes a new seasonal cycle which is in equilibrium with the boundary conditions and the external forcing, and interates. In this sense, Pollard assumes a state of quasi-equilibrium during the long term evolution of the climate.

Given a state of disequilibrium between the components of the AOC system, the evolution of the climate is critically dependent on the response time scales of the interacting components. For internal oscillations to occur with a period on the order of several thousand years, it is necessary that two components of the AOC system have similar response time scales which are on the order of several thousand years. Continental ice sheets have the required response time scale, and the ocean provides a second component with the required time scale if it is treated as a single isothermal heat reservoir. A number of workers used coupled AOC models in which they treated the ocean as a single isothermal reservoir having a single time constant, either on a global basis or in individual latitude zones (1,3,7,9). Sergin (10) also developed a coupled AOC model in which he modelled ocean heat transfer as a diffusion process and therefore was able to have a non-isothermal ocean, but he effectively replaced this model with an isothermal ocean having a single time constant.

In this paper we shall demonstrate that the ocean cannot be realistically treated as a single time constant component of the AOC system, but rather, responds with multiple time scales or with a continuously changing time scale. In particular, the initial surface temperature response to a radiative imbalance is governed by the relatively small mixed layer heat capacity rather than by the much larger deep ocean heat capacity, and is therefore relatively rapid. This rapid response is crucial in coupled AOC models, as the interaction between the atmosphere, cryosphere, and ocean surface determines the subsequent evolution of the climate. In the following section we begin by simply subdividing the isothermal ocean reservoir used in current AOC models into separate layers or boxes representing the mixed layer and deeper ocean layers, and show how the surface ocean temperature response varies with the number of layers and the mass exchange rates between these layers. Then we proceed one step further and replace the isothermal box model with a globally averaged one-dimensional advection-diffusion model of the deep ocean, and compare the ocean temperature response obtained with this model with that obtained with the box model and single ocean reservoir models. Finally, in order to illustrate the effect of ocean model formulation on the solution to simple AOC models, we take the model of (7) and replace their isothermal ocean reservoir with our box and advection-diffusion models and investigate the consequences for internally-generated oscillatory solutions.

However, our models employ highly simplified parameterizations of heterogeneous regional mixing processes. These processes include Ekman pumping, double diffusion, and convective overturning. We have parameterized these regional processes in terms of globally averaged processes of advection and diffusion, neither of which may be appropriate in the global average. It is possible to obtain the correct equilibrium simulation of, for example, the globally averaged vertical ocean temperature profile, but to obtain the wrong sensitivity to a perturbation using globally averaged parameterizations. We therefore perform a large number of sensitivity studies in an effort to bracket the likely response characteristics of the coupled AOC system in nature.

TRANSIENT RESPONSE OF SIMPLE BOX MODELS

Model description

The first model we use is a globally averaged box model consisting of an atmosphere layer and an ocean mixed layer, intermediate layer, and bottom layer. Heat exchange between the atmosphere and mixed layer occurs by radiative, latent, and sensible heat fluxes, and heat exchange between the ocean layers occurs through mass exchanges which represent the thermohaline circulation and Ekman pumping. Further details are provided in Harvey and Schneider (4).

Transient response to a solar constant increase

The equations governing the atmosphere and mixed layer can be represented as

$$R \frac{dT}{dt} = Q(t) - \lambda T + \text{(higher order terms in T)} \quad [1]$$

where $Q^*(t)$ is the absorbed solar radiation, λ is a damping coefficient which depends on the linearized radiative and albedo temperature feedbacks, T is temperature, and R is the system heat capacity. The infrared emission to space is commonly parameterized as $L_{OUT} = A + BT$ in energy balance models, and if the planetary albedo varies as $\alpha = a + bT$, then $\lambda = B + bQ$. A general solution of Eq.[1] is of the form $T(t) = T_q + T_{tr}(t)$, where T_q is the equilibrium solution and $T_{tr}(t)$ is the transient response and is of the form,

$$T_{tr}(t) \propto e^{-t/\tau} \quad [2]$$

where the response time scale or "e-folding" time τ is given by $\tau = R/\lambda$. After a time τ the climate warms by $1 - 1/e$ (about 66%) of its equilibrium warming. Note that this time scale, and hence the overall transient response, depends on both the thermal inertia R and the damping coefficient λ. For our model $\lambda = 1.93$ W m^{-2} K^{-1}, which is about the median value for state-of--the-art climate models.

The box model is numerically integrated until equilibrium (net radiation $<10^{-5}$ W m^{-2} in absolute value) is obtained with the present solar constant, then a 2% step function solar constant increase is applied. Figure 1 shows the transient atmosphere (dashed lines) and mixed layer (solid lines) temperature response for various versions of the ocean model. To avoid confusion, we show the atmosphere response only for two of the four cases presented in Figure 1, but the pattern of atmosphere warming is similar to the mixed layer warming in each case. Also shown as the horizontal dashed and solid lines are the atmosphere and surface warmings, respectively, which correspond to one e-folding time for a single time scale system. The atmosphere and mixed layer thermal inertias are 1.06×10^7 $Jm^{-2} K^{-1}$ and 1.23×10^8 $Jm^{-2} K^{-1}$, respectively, which yield response time scales of 2 months and 2 years. However, the atmosphere and surface are tighly coupled through radiative and turbulent fluxes and so respond jointly as a single system in the global average. Case of Fig. 1 shows the atmosphere and mixed layer responses when the mixed layer is decoupled from the deep ocean by setting the mass exchange rates to zero. As can be seen from Fig. 1, the e-folding time when the mixed layer is the only ocean component is 2.2 years, which is a simple sum of the individual atmosphere and mixed layer time scales.

Case B shows the atmosphere and mixed layer responses when the intermediate layer is coupled to the mixed layer with a mixed layer mass turnover time of 10 years. This case shall be referred to as the three box model. Also shown for this case is the intermediate layer response, which is about two orders of magnitude slower than the mixed layer response. Case C shows the responses for all the model components when both the intermediate layer and bottom layer are coupled to the mixed layer, yielding a four box model. Neither deep ocean component has much effect on the mixed layer during the first half of the equilibrium mixed layer response. During this period the mixed layer responds with a time scale governed by the mixed layer thermal inertia. Thereafter, the mixed layer responds with a time scale governed by the deep ocean thermal inertia. The change in surface response time scales from that of the mixed layer to that of the deep ocean occurs when the heat flux to the deep ocean approximately balances the net radiation at the top of the atmosphere. In equilibrium there is no heat flux to the deep ocean, but as the mixed layer warms in response to a solar constant increase a heat flux to the deep ocean arises and grows as the mixed layer - deep ocean temperature difference grows. Once approximate balance between the net radiation and deep ocean heat flux is obtained, subsequent atmosphere-surface warming is very slow and dependent on a gradual deep ocean warming, which slowly reduces the deep ocean heat flux.

Case D shows the surface and atmosphere temperature responses when we combine the mixed layer, intermediate layer, and bottom layer into a single isothermal ocean reservoir. This case corresponds to the ocean model assumption which (1,3,7,9) used in their coupled AOC models. In this case the temperature response is dramatically different from any of the three previous cases. The surface temperature now responds from the beginning with a time scale governed by that of the global ocean volume, whereas for the four box model the first 2/3 of the surface temperature response was governed by the mixed layer

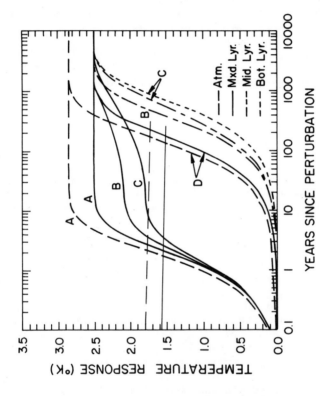

Figure 1 Box model transient response. The solid lines show the mixed layer tempeature response, the dashed lines show the atmospheric temperature response, the dash-dot lines show the bottom layer response. Case A : two box model (atmosphere and mixed layer). Case B : three box model (two box model and intermediate layer). Case C : four box model (three box model and bottom layer). Case D : single isothermal ocean reservoir. The dashed horizontal line corresponds to a warming of $(1-1/e)$ of the equilibrium atmospheric warming, and the solid horizontal line gives the same for the mixed layer. For single time constant systems, the intersection of these lines with the temperature graph gives the e-folding time τ.

thermal inertia. The e-folding time scale for this combined ocean boxes case is 183 years and is almost two orders of magnitude greater than the initial time scale for the four box model. As we shortly discuss, the difference in response time scales between the four box and the single ocean reservoir models is crucial to the existence of internal oscillations.

As shown in (4), the mixed layer response of the four box model is very insensitive to large changes in the mass exchange rates between the ocean layers in our model. Hence, our previous statement that the ocean surface temperature responds with more then one time scale remains valid despite large uncertainties in the mass exchange rate between the mixed layer and deeper ocean.

TRANSIENT RESPONSE OF A BOX ADVECTION DIFFUSION MODEL

We have used the simplest model which still retains the vital distinction between the mixed layer and the deep ocean. The simplicity of this model is appealing since computational requirements can easily become a limiting factor for long climate simulations. Although the box model encorporates an essential element of the ocean thermal response, it is admittedly very crude. In this section we compare the box model results with the results of a box-advection-diffusion model (BAD model). Although this model is slightly more sophisticated than the box model, it still contains a globally averaged parameterization of many heterogeneous regional physical processes. The model is described in detail in (4) and was recently employed by Hoffert, Callegari and Hsieh (6). The mixed layer is coupled to the deep ocean through a thermohaline advection and by thermal diffusion. Water leaves the mixed layer at a temperature of bottom water formation θ_B, sinks to the ocean bottom, and then upwells uniformly over the world ocean to return to the mixed layer. The effects of vertical mixing processes and Ekman pumping are represented by diffusion. For our base case we use a thermal diffusivity $k = 6.0 \times 10^{-5}$ m^2 s^{-1}, an advection velocity $w = 4$ m yr^{-1}, and a temperature θ_B of bottom water formation of 1.2°C, taken from (6), although the appropriate globally averaged values of k and w are highly uncertain.

Figure 2 shows the transient mixed layer response and mean deep ocean warming for the BAD model with the base case values of k and w (Case A) given above, as well as the response when k and w are both halved (Case B), both doubled (Case C), and when w is set equal to zero (Case D)[3]. We also show as cases E and F the four box model and single world ocean reservoir results (Cases C and D of Fig. 1). Comparison of cases A, B and C indicates that the mixed layer response is relatively insensitive to

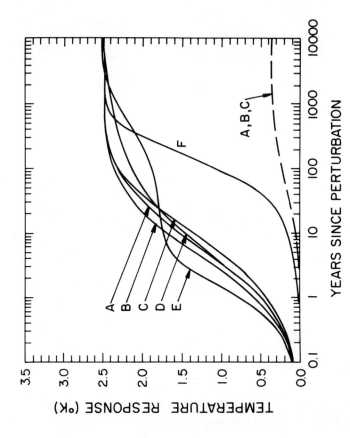

Figure 2 Sensitivity of the transient response of the BAD model and comparison with the four box and single ocean reservoir models. The solide lines show the mean deep ocean warming for the BAD model. Case A : base case, values of k and w as given in text. Case B : k and w are two times base case values. Case C : k and w are half the base case values. Case D : k = base case value, w = 0. Case E : four box model base case. Case F : single isothermal world ocean reservoir.

effect of ocean thermal response formulation with the KCG model simply because we found that all the parameters required to reproduce their results were explicitly given. In implementing our ocean models in the KCG model, we simply add the appropriate deep ocean energy flux terms to their surface termperature equation, and add the appropriate deep ocean equations (eq.[10] and 11 of (4) for the box model, and eq.[13] of (4) for the BAD model). We retain their glacier growth equation, and compute all the surface energy fluxes as in their model rather than as in our models. The resulting system of equations is numerically integrated using the 5th order Runge-Kutta subroutine DVERK, which is part of the IMSL package.

As a check on our computer code and in order to gain some insight into the behaviour of the KCG model, we perform a sensitivity study of their model by systematically varying some of their model parameters, one at a time. The parameters which are varied and the range of values for which we obtain bounded oscillations are shown in Figure 3. At one end of the range for each parameter there are no oscillations, and as the parameter is moved to the other end of the range oscillations develop and increase in amplitude until a point is reached where the model undergoes a catastrophic cooling.

In Figures 4 and 5 we show the response for the base case of the KCG model, but with their isothermal ocean replaced by our box model and BAD model ocean, respectively. In each case the model mixed layer underwent a catastrophic cooling to about 200°K, at which point glacier retreat had commenced and the model climate started to recover. In the case of the box model a negative argument arose in a square root function and the simulation aborted, while for the BAD model an equilibrium state with zero ice extent was reached at a mixed layer temperature of 310°K.

In the original KCG ocean model it is the slow ocean surface temperature response to a radiative imbalance which allows reasonably bounded internal oscillations to occur. For example, if the surface temperature responds to a radiative deficit with a time scale comparable to the glacier response time scale, then a large cooling will not have occurred before the glacier begins to shrink as a result of smaller snow accumulation (following, for the sake of argument, the reasoning behind the KCG model). The smaller glacier size in turn will reduce the radiative deficit and slow down the rate of cooling. If, on the other hand, the surface temperature response is fast compared to the glacier response, then a catastrophic cooling occurs before significant glacier retreat can occur (again assuming that glaciers retreat as the climate cools). It is worth emphasizing the fact that most AOC models postulate a continuous state of disequilibrium

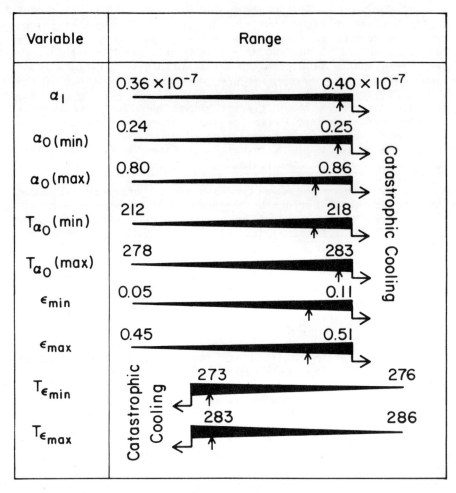

Figure 3 Sensitivity of the KCG model (7) to parameter changes. $\alpha_1 = d\alpha_L/dL$, where α_L = land albedo and L = glacier extent. α_{0min} = minimum ocean albedo. $\alpha_{0(max)}$ = maximum ocean albedo. $T_{\alpha_0(max)}$ = temperature of maximum ocean albedo. ε_{min} = minimum glacier mass accumulation/ablation ratio. ε_{max} = maximum glacier mass accumulation ratio. $T_{\varepsilon min}$ = temperature of minimum accumulation/ablation ratio. $T_{\varepsilon max}$ = temperature of maximum accumulation/ablation ratio. See (7) for further explanation. The width of the horizontal lines is proportional to the amplitude of the climate oscillations, the numbers indicate the range of the corresponding parameter, and the arrows indicate the parameter values used in (7).

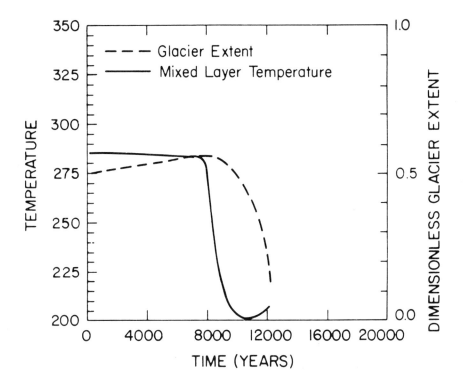

Figure 4 Time dependent mixed layer temperature and glacier extent for the KCG model when their isothermal world ocean reservoir is subdivided into a mixed layer, intermediate layer, and bottom layer as in our box model.

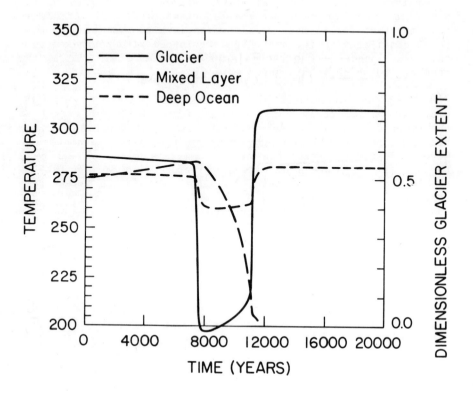

Figure 5 Time dependent glacier extent and mixed layer and mean deep ocean temperature for the KCG model using a box advection-diffusion ocean model.

between the various components as the climate evolves, and it is for this reason that the different response time scales of the mixed layer and deep ocean are important. This difference is not important for climate models which involve a continuous state of quasi-equilibrium and which are forced with periods longer than the longest ocean thermal response time scale, as in the case of (8) which we discussed earlier.

Given the difference in response time scales for the glacier extent and ocean surface temperature which now exists in the KCG model, there is no reason to expect that we can recover the original oscillations by varying the model parameters. Nevertheless, we varied the nine parameters given in Figure 3 through the same range of values as before, but were unable to obtain oscillations. Instead, the model climate approached temperatures near 310°K or near 275°K which were maintained to the end of the 100 kyr integration periods which we used, or else underwent a castrophic cooling. These results raise the possibility of a potentially interesting intransitivity.

CONCLUDING COMMENTS

We have demonstrated the importance of distinguishing between the fast responding mixed layer and the slower responding deep ocean when building coupled atmosphere-ocean-cryosphere (AOC) models. The AOC models exhibiting internal oscillations which we have examined here rely on a slow ocean surface temperature response to a radiative imbalance, which is obtained by treating the ocean as a single isothermal reservoir. We have shown using a series of simple box models and using a one dimensional advection-diffusion deep ocean model with widely varying parameter values that most of the surface temperature response occurs much more rapidly than that of the deep ocean. It is only the final approach to equilibrium that occurs with a time scale characteristic of the deep ocean.

What previous AOC models have demonstrated is that internal oscillations can be solutions to coupled atmosphere--ocean-cryosphere models. But, the most difficult problem will be to demonstrate the extent to which their simple formulation of complex processes is realistic, and even if realistic, that the numerical values of key parameters are also realistic. To make such a demonstration we need to compare results of sensitivity experiments across a hierarchy of coupled models of increasing complexity and to compare these results with the available data. At present, the state-of-the-art cannot rule out the possible existence of internal oscillations on Pleistocene time scales. But, because of the extreme sensitivity of the solutions to plausible internal model characteristics, neither

the values of k and w. This insensitivity is fortunate given the large uncertainty in the values of k and w. Comparison of cases A and D (where w=0) indicates that the diffusion term is the most important of the two deep ocean terms in controlling the mixed layer response.

In (5) we compare the first 15 years of the transient response of a version of the BAD model having land-sea resolution with the transient response of the coupled atmosphere-ocean general circulation model of (2). We find that the transient responses for the two models differ by about 0.1 to 0.2°C at any given time. Hence, despite the use of globally averaged parameterizations, our model transient response is reasonably close to that of the detailed general circulation model in (2).

The mixed layer response for the BAD model is clearly closer to the four box model response than to the single ocean reservoir response, although it lacks the two distinct time scales which characterize the four box model response. As we explain in (4), the mixed layer responds with a continuously increasing time scale. Nevertheless, the most important feature of the four box model, namely, the initially rapid mixed layer response, is retained in the BAD model.

An interesting feature of the BAD model is that the mean deep ocean warming is only a small fraction of the mixed layer warming for the version of the BAD model presented here, whereas in the box model the deep ocean is forced to warm by the same amount as the mixed layer warming. A small mean deep ocean warming occurs because k, w and θ_B are constant here. In (4) we allow a variety of plausible feedbacks to occur between these parameters and the mean mixed layer temperature, and show that a mean deep ocean temperature change ranging from a warming much larger than the mixed layer warming to a cooling can occur as the mixed layer warms. The point we wish to emphasize here, however, is that the mean deep ocean warming is not constrained to be equal to the mixed layer warming for the BAD model, whereas it is constrained in the case of the box model. This distinction would be particularly important to the Saltzman and Moritz AOC model (9), in which the atmosphere CO_2 concentration depends on the temperature of an isothermal global ocean reservoir.

EFFECT OF OCEAN MODEL FORMULATION ON AOC MODELS

In order to test the effect of ocean model formulation on AOC models, we have implemented the box ocean model and the BAD ocean model in the coupled AOC model of (7), which we shall henceforth refer to as KCG. We have chosen to illustrate the

has the state-of-the-art provided any convincing evidence for such oscillations in reality.

ACKNOWLEDGEMENT

LDDH received financial support from a Natural Science and Engineering Research Council of Canada grant to F.K. Hare.

REFERENCES

1. Bhattacharya, K., Ghil, M., and Vulis, I.L. : 1982, J. Atmos. Sci. 39, pp. 1747-1773.
2. Bryan, K., Komro, F.G., Manabe, S., and Spelman, M.H. : 1982, Science 215, pp. 56-58.
3. Ghil, M., and Le Treut, H. : 1981, J. Geophys. Res. 86, pp. 5262-5270.
4. Harvey, L.D.D., and Schneider, S.H. : 1984a, J. Geophys. Res. (in press).
5. Harvey, L.D.D., and Schneider, S.H. : 1984b, J. Geophys. Res. (submitted).
6. Hoffert, M.I., Callegari, A.J., and Hsieh, C.T. : 1980, J. Geophys. Res. 85, pp. 6667-6679.
7. Kallen, E., Crawford, G., and Ghil, M. : 1979, J. Atmos. Sci. 36, pp. 2292-2303.
8. Pollard, D. : 1982, Climate Research Institute, Report n°37, Oregon State University, Corvallis, Oregon, 43 pages.
9. Saltzman, B., and Moritz, R.E. : 1980, Tellus 32, pp. 93-118.
10. Sergin, V.Ya. : 1979, J. Geophys. Res. 84, pp. 3191-3204.

[1] Also affiliated with Department of geography, University of Toronto, Toronto, Canada M5S 1A1.
[2] The National Center for Atmospheric Research is sponsored by the National Science Foundation.
[3] We vary K and w together rather than one at a time in order to preserve the same ratio K/w, which determines the shape of the deep ocean temperature profile and is therefore rather well constrained. See (4) for the effects of choosing K and w individually.

PART III

MODELLING LONG-TERM CLIMATIC VARIATIONS IN RESPONSE TO ASTRONOMICAL FORCING

SECTION 4 – CONCEPTUAL MODELS OF CLIMATIC RESPONSE

AN EVALUATION OF OCEAN-CLIMATE THEORIES ON THE NORTH ATLANTIC

W.F. Ruddiman[1] and A. McIntyre[1,2]

[1]Lamont-Doherty Geological Observatory of Columbia University, Palisades, New York
[2]Queens College of the City University of New York, Flushing, New York

OCEAN-CLIMATE THEORIES

The role of the oceans in ice-age climatic change has been conceived differently in various publications over the last two decades. Figure 1 summarizes three broad categories into which these viewpoints might be divided. The focus here is on the role of the mid-latitude to high-latitude North Atlantic Ocean.

One group hypothesizes that the ocean can initiate major changes in glaciation on adjacent land masses by means of the early response of its high-latitude sea-ice cover. Early sea-ice advances lead sequentially to higher Northern Hemisphere albedo, lower atmospheric temperatures, and thus the initiation of glaciation. This group of hypotheses, which we will call the "albedo initiator" school, may look to atmospheric or other forcing in geographically removed areas to initiate the sea-ice advance, but has the common attribute that the local oceanic response (in the North Atlantic) precedes and triggers the nearby land/ice sheet response (1-3). The converse sequence occurs during periods of ice decay.

A second group contends that the ocean plays an active role in triggering climatic change through variations in the flux of moisture to high-latitude land masses. Early increases in sea-surface temperature (SST) and decreases on sea-ice cover enhance the uptake of latent heat by the atmosphere and thus increase precipitation on land and initiate glaciation (Fig. 1). We will call this the "moisture initiator" school. A prominent early version of this class of hypothesis emphasized the Arctic

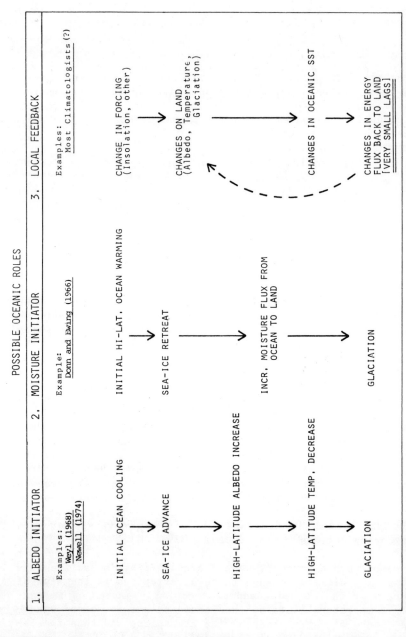

Figure 1 Possible roles played by the ocean in climatic change. All three categories show sequences keyed to the ice-growth path of glacial cycles; converse-sense sequences would occur on ice decay.

Ocean as a moisture source (4), but other versions have generally looked to more energetic sources of latent heat in the Gulf of Mexico and North Atlantic (3, 5-7). Several papers have emphasized the importance of low- or mid-latitude moisture sources without formulating a complete climatic hypothesis (8-11).

The first two groups of climatic theories in Figure 1 arrive at glaciations from exactly opposite initial assumptions (oceanic warmings and coolings). This occurs in part because each group focuses on only one of the components of the air-sea energy exchange: sensible heat in the albedo initiator hypotheses and latent heat in the moisture initiator hypotheses. In fact, the effects of these two energy components operating simultaneously obviously could tend to counteract each other on a local scale: for example, increased flux of moisture in a given region may enhance ice accumulation, while increased flux of sensible heat would simultaneously enhance ice ablation. In general, paleoclimatologists are open to the criticism of only focusing on those portion of the climatic system à priori postulated as critical, while ignoring other components.

A third, less easily traced viewpoint is that which would probably be taken by most climatologists, based on their familiarity with the modern climate system. The ocean at middle and high latitudes is viewed as a subsidiary part of the climatic system, one which moderates geographic extremes by transporting heat poleward and moderates seasonal extremes by its large thermal inertia. The short-term effect of the atmosphere upon the surface ocean is both distinct and rapid, with a characteristic response time of months to at most years, excluding the much slower response possible in the deep ocean. The wind-driven response of the surface ocean to atmospheric forcing is clearly defined and rapid, and the marked seasonal response of mid-latitude SST to radiative forcing and continental air masses (with appropriate lags of roughly 1-2 months) implies a rapid, but essentially passive thermal role for the oceans. Furthermore, at high to middle latitudes, it is difficult to demonstrate conclusively from modern observations that oceanic SST anomalies have any precursor impact upon atmospheric circulation (12). As a result, most climatologists would see the mid-latitude to high-latitude ocean primarily as a reactive, rather than an independently active, part of the climate system.

Even if the North Atlantic were proven to be primarily reactive to other forcing, it would still remain important for ice-age climatology whether the feedback it gives to the ice sheets is large or small, and positive or negative. Much of the criticism during the 35 years that separated the initial formulation of Milankovitch's orbital theory (13) and its ultimate vindication by time series and spectral analysis of deep-sea records (14-15) took one form: changes in incident radiation of a few percent for the Milankovitch caloric seasons and ±1% or less annually at any one place on the earth's surface were thought insufficient to cause such large climatic responses (16-20). Even though the changes in the monthly insolations are much larger (up to 12%, (47)) and recognition of orbital frequencies in many kinds of deep-sea climatic records has confirmed the basic orbital control of glaciation, the need for feedback processes to amplify orbital effects remains. There is, however, substantial disagreement over which feedbacks are important. The ocean is a plausible source of significant energy feedback.

NEW EVIDENCE

For paleoceanographic purposes, the most intensively studied region in the world ocean is the North Atlantic between 35°N and 65°N. This interest reflects its critical location adjacent to the ice-age ice sheets and astride regions of intermediate and deep-water formation. Studies through the mid-1970's laid out the basic late Quaternary patterns of faunal and floral change (21-24) and of estimated SST variation (25). The major result of these efforts was the discovery that a mass of frigid, ice-laden polar water periodically filled the subpolar North Atlantic southward to a sharply defined front at 45°N during glaciations.

More recent work (26, 27) has dealt with the spectral character and phasing of these surface-ocean changes, especially in comparison to that of the $\delta^{18}O$ ($\sim$ice-volume) record. The results summarized here derive entirely from these two most recent studies.

Spectral analysis of SST time series covering almost 250 kyr and derived from North Atlantic cores between 41°N and 56°N has shown several features schematically summarized in the top of Figure 2 : (i) dominant, or very strong, 100 kyr power in all cores, although the character of the 100 kyr signal appears to change at about 45°N; (ii) 23 kyr power dominant in core V30-97 at 41°N, and declining rapidly northward to lower values at 54°N; and (iii) 41 kyr power at a minimum in the core at 41°N but dramatically larger in the three cores between 50° and 54°N. These patterns indicate a very large change in the spectral response of North Atlantic surface waters over very short distances.

These results show that the cold "polar-water" gyre that periodically developed north of 45°N fluctuated primarily with a 100 kyr and 41 kyr rythm, whereas a very different 23 kyr periodicity characterized the response of the northern subtropical gyre immediately to the south. This change in character of the SST spectra occurs in the middle of the region of largest ice-age thermal response (Fig. 2, top). The diminished response south of 35°N is associated with the stable center of the subtropical gyre (29); north of 65°N, the Norwegian Sea

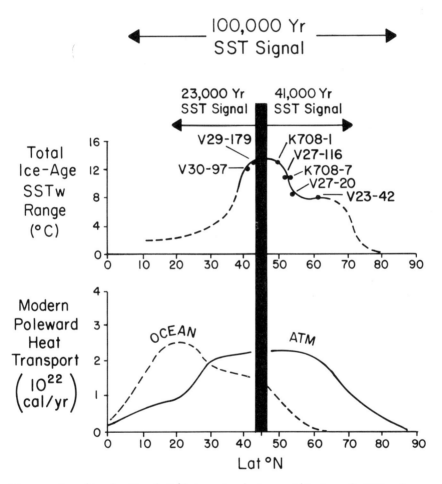

Figure 2 (top) Glacial/interglacial amplitude of SST change and geographic partitioning of frequency response (from 27); (bottom) Estimate of modern poleward transport of heat by the Northern Hemisphere oceans and atmosphere (from 28).

waters remained almost constantly frigid, with more temperate water penetrating from the south only every 100 kyr or so during peak inter-glaciations like today (30, 31).

The discussion of SST/ice volume phase relationships that follows is built on two assumptions: (i) that the $\delta^{18}O$ signal largely reflects ice volume, despite complicating local overprints due to temperature and other factors (32); (ii) that the

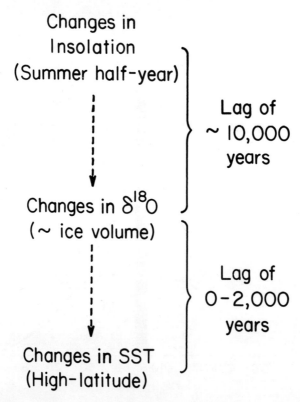

Figure 3 Phasing of the high-latitude 41 kyr North Atlantic SST signal relative to high-latitude Northern Hemisphere summer insolation and high-latitude ice volume, both at the 41 kyr period ; based on results in (27).

presence of 41 kyr and 23 kyr periodicities in the $\delta^{18}O$ record reflects geographic partitioning of the ice-volume response into, respectively, a high-latitude and mid-latitude component. The latter assumption derives from the similar geographic partitioning of insolation forcing (13, 33).

Analysis of phase relationships shows that the filtered 41 kyr SST signal north of 50°N is, within the uncertainty due to sampling control, in phase with the same periodicities in the high-latitude $\delta^{18}O$ (~ice volume) signal (Fig. 3). This approxi-

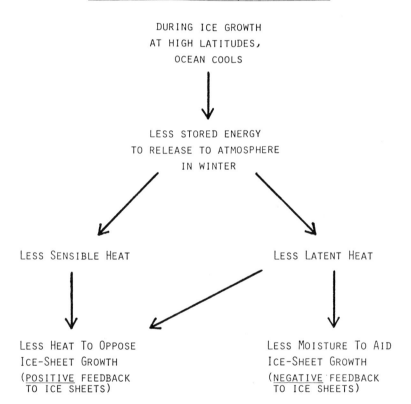

Figure 4 Partinioning of the high-latitude North Atlantic energy feedback to high-latitude ice sheets at the 41 kyr period into heat and moisture feedback components.

mately synchronous phasing implies that the high-latitude ice sheets guide the circum-Arctic regions into and out of glacial conditions by means of albedo feedback. This suggests a primarily reactive role for the ocean at latitudes north of 50°N and at the 41 kyr and 100 kyr frequencies.

The $\delta^{18}O$/SST phasing at the 41 kyr cycle indicates that the high-latitude ocean provides positive feedback to the insolation-driven oscillations in the ice sheets through changes in oceanic heat flux but negative feedback through changes in the moisture flux (Fig. 4). Because the ocean north of 50°N carries only modest amounts of the total hemispheric heat flux (Fig. 2, bottom), its role in the total global heat transfer appears to be minor; however, proximity to the ice sheets leaves open the possibility that this modest transfer could be significant to ice-sheet mass balance. In addition, because ice shelves, sea ice, and marine-based ice sheets are in direct contact with the ocean, cooling of the sea by heat extraction may be the single most critical feedback process for high-latitude ice mass balance.

The 23 kyr SST signal that is so strong south of 50°N lags roughly 6 kyr (one-quarter wave length) behind the 23 kyr $\delta^{18}O$ signal (Fig. 5). This phasing has been attributed to two major factors: (i) the effects of meltproducts from the mid-latitude ice sheets, which fluctuate in size with the 23 kyr periodicity; and (ii) the impact of variable advection of warm water northward from the low latitudes of the North and South Atlantic (27, 34).

Because the 23 kyr SST period dominates in the highly energetic circulation of the northern subtropical gyre (Fig. 2, bottom), oceanic feedback processes at this periodicity may be very significant both locally and globally. This part of the ocean provides a negative feedback of heat to ice-sheet oscillations at the 23 kyr periodicity, but a positive moisture feedback that is potentially very powerful, especially in winter (Fig. 6). We specify winter because at these latitudes it is the major season of moisture extraction from the ocean. The largest latent-heat fluxes in the world ocean occur at mid-latitudes in winter off the east coasts of North America and Asia (35, 36).

The separate, frequency-specific spectral responses and SST/$\delta^{18}O$ lags for the high-latitude North Atlantic (Fig. 4) and the mid-latitude region (Fig. 6) are in fact blended across a transitional zone reaching from about 45°N to 55°N (Fig. 2). In addition, there have been intervals in the past during which the well-marked oceanic lag characteristic of the mid-latitudes

INSOLATION/ICE/OCEAN PHASING AT THE 23,000-YEAR (PRECESSIONAL) CYCLE

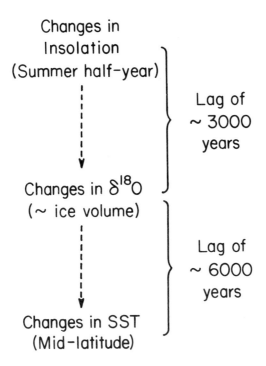

Figure 5 Phasing of the mid-latitude 23 kyr North Atlantic SST signal relative to mid-latitude Northern Hemisphere summer insolation and mid-latitude ice volume, both at the 23 kyr period ; based on results in (27).

has penetrated to the north. For example, the North Atlantic as far north as 55° remained warm during several thousand years of ice growth at the stage 5/4 and 5e/5d boundaries (37, 38) and remained cold for a comparable part of the stage 2/1 deglaciation (39). Apparently, the geographic changes of SST spectral power and the associated energy feedbacks shown in Figures 2 to 6 represent the mean partitioning of the North Atlantic response over the last 250 kyr, but may overlook

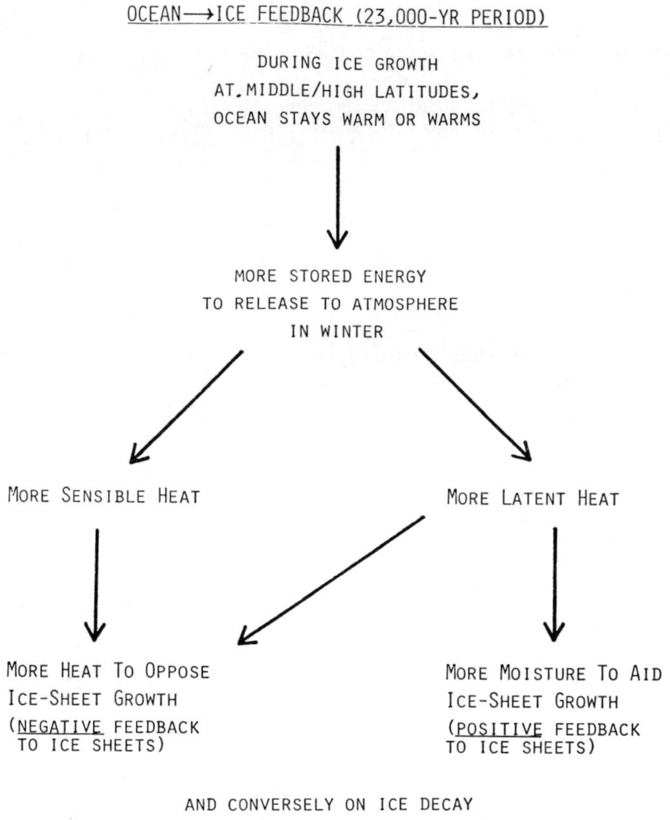

OCEAN→ICE FEEDBACK (23,000-YR PERIOD)

DURING ICE GROWTH
AT MIDDLE/HIGH LATITUDES,
OCEAN STAYS WARM OR WARMS

↓

MORE STORED ENERGY
TO RELEASE TO ATMOSPHERE
IN WINTER

↙ ↘

MORE SENSIBLE HEAT   MORE LATENT HEAT

↓   ↓

MORE HEAT TO OPPOSE   MORE MOISTURE TO AID
ICE-SHEET GROWTH   ICE-SHEET GROWTH
(NEGATIVE FEEDBACK   (POSITIVE FEEDBACK
TO ICE SHEETS)   TO ICE SHEETS)

AND CONVERSELY ON ICE DECAY

Figure 6 Partitioning of the mid-latitude North Atlantic energy release to mid-latitude ice sheet at the 23 kyr period into heat and moisture feedback components.

several potentially critical northward movements in the "lagging-ocean" mid-latitude regime during large climatic transitions. There may also be significant east-west differences that we cannot detect with our largely north-south core array.

ASSESSMENT OF OCEAN-CLIMATE THEORIES

Based on the new evidence just summarized, we show in Table 1 an assessment of the ocean-climate theories and hypotheses as they specially apply to the North Atlantic. Because the North Atlantic frequency response is regionally partitioned (Fig. 2, top), we specify in Table 1 the applicability of each theory/hypothesis at both the 41 kyr and the 23 kyr frequencies.

The evidence rules out any possibility that the North Atlantic between 41° and 56°N is an albedo initiator of glaciations. At neither the 41 kyr nor the 23 kyr frequency does the oceanic SST response actually lead $\delta^{18}O$/ice volume (Figs. 4, 6). Any plausible decoupling between $\delta^{18}O$ and true ice volume (40) could only increase the oceanic lags behind true ice volume.

We can also rule out the "moisture-initiator" role for the highest latitudes of the North Atlantic at the 41 kyr frequency, because the ocean there cools as high-latitude ice sheets grow. Only one possible way remains in which the North Atlantic could play an independent initiator role in glaciations: through the mid-latitude response at the 23 kyr period. If that signal is caused by cross-equatorial advection of heat, and if the winter moisture release from the ocean is more important than that of heat (Fig. 6), then the North Atlantic south of 50°-55°N could play an independent role as a climatic initiator at the 23 kyr periodicity.

At least two other published hypotheses invoked an independent "initiator" role for the ocean and also anticipated some aspects of the geographic partitioning of ocean-climate response developed here, but they were flawed in other respects. One hypothesis (5) called for a "polar" ocean with a sympathetic phase relationship to ice-sheet size and a mid-latitude ocean with the same kind of phase lag relative to ice as suggested here. However, this model of the polar ocean founders on the evidence against an ice-free Arctic during the last million years (41, 42). Furthermore, this hypothesis was not tied to orbital forcing; instead, it was assumed that the ice-ocean interactions that constituted the glacial cycles would be self-perpetuating.

Another hypothesis (3) called on orbital forcing to cause an early cooling and sea-ice advance in the Norwegian Sea/Northeast Atlantic that resulted in an oceanic warming in the Labrador Sea and a moisture initiated glaciation over North America. However, our evidence argues against any significant precursor response of the high-latitude North Atlantic, which followed with negligible lag the ice-volume response at the 41 kyr cycle.

As for the dependent feedback roles listed in Table 1, evidence summarized in the last section (Figs. 4, 6) shows that

TABLE 1. ASSESSMENT OF NORTH ATLANTIC ROLE IN GLACIAL CYCLES

	Independent Role: Climatic Initiator		Dependent Role: Local Feedback To Changes Initiated on Land	
	Albedo/Temp. Initiator	Moisture Initiator	Heat/Albedo Feedback	Moisture Feedback
41,000-YR PERIOD	NO	NO	YES (POSITIVE)	YES (NEGATIVE)*
23,000-YR PERIOD	NO	MAYBE*	YES** (NEGATIVE)	YES** (POSITIVE)

*Portion of 23,000-yr SST signal driven from low latitudes or by cross-equatorial flux.

**Portion of 23,000-yr SST signal driven by meltwater/iceberg influx, not by Albedo/Temp. changes on land.

the North Atlantic provides feedbacks of moisture and heat at the 41 kyr and 23 kyr periodicities, but the heat and moisture effects at each periodicity probably act in opposite senses in terms of their effects on glacial mass balance. For the reasons noted, we favor the positive feedbacks of moisture at the 23 kyr period and heat at the 41 kyr period as critical, but this assertion needs testing by climate models capable of dealing with the complexities of the air-sea-ice system.

Finally, there, is disagreement among glacial geologists as to whether high-latitude (circum-Arctic) and mid-latitude (Laurentide, Scandinavian) ice sheets reacted in phase or out of phase with each other (43). This disagreement hinges in part on the significance attributed to the sea-level calving mecanism (44, 45) and partly on the relative importance of temperature versus moisture as critical factors in glaciation.

Those emphasizing the importance of cold (summer) air temperatures tend to infer in-phase behavior of ice sheets throughout the Northern Hemisphere because of regionally synchronous refrigeration due to lower insolation and albedo feedback. The major mechanism at work in these theories is reduced ablation because of colder temperatures. In contrast, those favoring increased precipitation as the critical factor stress out-of-phase behavior, with high-Arctic ice sheets starved for moisture during the most extensive advances of mid-latitude ice sheets toward the major precipitation sources.

The fact that $\delta^{18}O$ variations occur at different orbital frequencies (15) negates any simple theory of either temperature or precipitation control. We suggest that the 41 kyr $\delta^{18}O$ component reflects ice fluctuations largely at latitudes poleward of 65° in both hemispheres, while the 23 kyr component reflects ice fluctuations at mid-latitudes of the Northern Hemisphere. Through time, $\delta^{18}O$ (~ice volume) minima and maxima are necessarily sometimes synchronous and sometimes opposed at the insolation signals at the two different frequencies, combined with the associated ice-response lags (15, 46). This discussion begs the question of the origin and geographic partitioning of the very large 100 kyr signal that dominates $\delta^{18}O$ (~ice volume) records.

In summary, neither strictly in-phase nor out-of-phase behavior of high- and mid-latitude ice sheets is observed, because neither temperature nor precipitation is the first cause of ice-age ice-sheet fluctuations. Orbitally controlled variations of insolation are the first cause of ice-sheet responses; the central issue of the oceanic role is its place in the feedback processes that amplify or suppress the effects of variable insolation forcing across different geographic regions. Our results show that the oceanic response is also geographically partitioned, and thus its feedback role is latitudinally variable as well.

ACKNOWLEDGMENTS

This research was supported by NSF grants ATM 80-19253s, OCE 80-18177, OCE 83-15237. This is also L-DGO contribution no. 3607.

REFERENCES

1. Weyl, P. : 1968, Meteoral. Monogr. 8, pp. 37-62.
2. Newell, R.E. : 1974, Quaternary Research 4, pp. 117-127.
3. Johnson, R.G., and McClure, B.T. : 1976, Quaternary Research 6, pp. 325-353.
4. Ewing, M., and Donn, W.L. : 1956, Science 23, pp. 1061-1066.
5. Stokes, W.L.: 1955, Science 122, pp. 815-821.
6. Emiliani, C., and Geiss, J. : 1957, Geologische Rundschau 46, pp.   576-601.
7. Donn, W.L., and Ewing, M. : 1966, Science 152, pp. 1706-1712.
8. Lamb, H.H., and Woodroffe, A. : 1970, Quaternary Research 1, pp. 29-58.
9. Chappell, J. : 1974, Nature 252, pp. 199-202.
10. Barry, R.G., Andrews, J.T., and Mahaffy, M.A. : 1975, Science   190, pp. 979-981.
11. Adam, D.P. : 1975, Quaternary Research 5, pp. 161-171.
12. Namias, J., and Cayan, D.R. : 1981, Science 214, pp. 869-876.
13. Milankovitch, M.M. : 1941, "Canon of Insolation and the Ice-Age Problem", Koniglich Serbische Akademie, Beograd. English Translations, published for the United States Department of Commerce and the National Science Foundation, Washington, D.C.
14. Broecker, W.S. : 1966, Science 151, pp. 299-304.
15. Hays, J.D., Imbrie, J., and Shackelton, N.J. : 1976, Science 194,   pp. 1121-1131.
16. Van Woerkom, A.J.J. : 1953, "Climatic Change", H. Shapley (Ed.),   Harvard University Press, Cambridge, pp. 147-157.
17. Simpson, G.C. : 1957, Roy. Met. Soc. Quart. Jour. 83, pp. 459-485.
18. Shaw, D.M., and Donn, W.L. : 1968, Science 162, pp.459-485.
19. Budyko, M.I. : 1969, Tellus 21, pp. 611-619.
20. Sellers, W.D. : 1969, J. Appl. Meteor. 8, pp. 392-400.

21. McIntyre, A., Ruddiman, W.F., and Jantzen, R. : 1972, Deep-Sea Research 19, pp. 61-77.
22. Ruddiman, W.F., and McIntyre, A. : 1973, Quaternary Research 3, pp. 117-130.
23. McIntyre, A., Kipp, N., Bé, A.W.H., Crowley, T., Gardner, J.V., Prell, W.L., and Ruddiman, W.F. : 1976, in "Investigation of Late Quaternary Paleoceanography and Paleoclimatology", R.M. Cline and J.D. Hays (Eds.), The Geological Society of America, Boulder, Geological Society of America Memoir, 145, pp. 43-76.
24. Ruddiman, W.F., and McIntyre, A. : 1976, in "Investigation of Late Quaternary Paleoceanography and Paleoclimatology", R.M. Cline and J.D. Hays (Eds.), The Geological Society of America, Boulder, Geological Society of America Memoir, 145, pp. 199-214.
25. Sancetta, C., Imbrie, J., and Kipp, N.G. : 1973, Quaternary Research 3, pp. 110-116.
26. Ruddiman, W.F., and McIntyre, A. : 1981, Science 212, pp. 617-627.
27. Ruddiman, W.F., and McIntyre, A. : 1983, Geol. Soc. of America Bull., in press.
28. Vonder Haar, T.H., and Oort, A.H. : 1973, J. Phys., Ocean 3, pp. 169-172.
29. Crowley, T.J.: 1981, Mar. Micropal. 6, pp. 97-129.
30. Kellogg, T.B.: 1975, "Climate of the Arctic", G. Weller and S.A. Bowling (Eds.), University of Alaska Press, Fairbanks, pp. 3-36.
31. Kellogg, T.B. : 1976, in "Investigation of Late Quaternary Paleoceanography and Paleoclimatology", R.M. Cline and J.D. Hays (Eds.), Geol. Soc. of America, Boulder, Geol. Soc. of America Memoir, 145, pp. 77-100.
32. Shackleton, N.J., and Opdyke, N.D. : 1973, Quaternary Research 3, pp. 39-55.
33. Berger, A.L. : 1978, Quaternary Research 9, pp. 139-167.
34. McIntyre, A., Mix, A.C., and Ruddiman, W.F. : in prep.
35. Bunker, A.F. : 1976, Mon. Weath. Rev. 104, pp. 1122-1140.
36. Budyko, M.I. : 1978, "Climatic Change", J. Gribbin (Ed.), Cambridge University Press, Cambridge, pp. 85-113.
37. Ruddiman, W.F., and McIntyre, A. : 1979, Science 204, pp. 173-175.
38. Ruddiman, W.F., McIntyre, A., Niebler-Hunt, V., and Durazzi, J.T. : 1980, Quaternary Research 13, pp. 33-64.
39. Ruddiman, W;F., and McIntyre, A. : 1981, Palaeogeography, -climatology, -ecology 35, pp. 145-214.
40. Mix, A.C., and Ruddiman, W.F. : 1983, Quaternary Research in prep.
41. Hunkins, K., Bé, A.W.H., Opdyke, N.D., and Mathieu, G. : 1971, in "Late Cenozoic Glacial Ages", K.K. Turekian (Ed.), Yale Univ. Press, New Haven, pp. 215-237.

42. Clark, D.L., Whitman, R.R., Morgan, K.A., and Mackey, S.D. : 1980, Geol. Soc. of Amer. Spec. Paper 181, pp. 57.
43. Boulton, G.S. : 1979, Boreas 8, pp. 373-395.
44. Hughes, T., Denton, G., and Grosswald, M. : 1977, Nature 266, pp. 596-602.
45. Denton, G.H., and Hughes, T. : 1981, "The Last Great Ice Sheets", Wiley Interscience Press, New York.
46. Imbrie, J., and Imbrie, J.Z. : 1980, Science 207, pp. 943-953.
47. Berger, A.L. : 1979, Il Nuovo Cimento 2C(1), pp. 63-87.

TERMINATIONS[1]

W. S. Broecker

Lamont-Doherty Geological Observatory of Columbia
University, Palisades, N.Y. 10964, USA

ABSTRACT

The coincidence between the termination of major glaciations in the northern and southern hemispheres is not easily explained. The association between these events and peaks in northern hemisphere seasonality suggests that these rapid warmings are linked to the earth's orbital cycles. The lack of correlation between terminations and the seasonality record for the southern hemispheres suggests that the southern hemisphere terminations must have been driven by events in the nothern hemisphere. The global sea level rise associated with the melting of the large northern hemisphere ice sheets does not appear to be the link because the events in the Antarctic Ocean clearly led the O^{18} change in the sea. Also, it is difficult to see why a sea level rise would trigger deglaciation in the Andes. Albedo change due to the retreat of the northern ice sheets and sea ice not only suffers the timing problem associated with the sea level change but appears to be inadequate on theoretical grounds. GCM models suggest that the radiation budget influence of the albedo changes would be confined to the northern hemisphere. Thus, we are left with either large changes in the pattern of deep sea ventilation or CO_2 changes in the atmosphere as the agents of interhemispheric transmittal of the climatic message.

INTRODUCTION

For more than two decades I have been a convert of Milankovitch. Ever since my graduate student days when Emiliani's (1) first O^{18} records from deep sea sediments appeared, thoughts regarding the exact connection between the seasonality record and the northern hemisphere ice record have bounced through my head. As we all know, the evidence for this connection has become ever stronger until now it is overwhelming. Indeed the seasonality changes produced by the precession of the orbit and the variations in the tilt of its axis do influence climate. Of late more confident statements are heard which carry the situation a step further claiming that seasonality changes are the cause of the primary 100 000 year glacial cycle as well as of the 21 000 and 41 000 year cycles which modulate this greater cycle. Were I to have place a bet as to whether this claim that Milankovitch does it all is correct, I suspect I would wager that it is. There is, however, one major puzzle in this regard which causes me to hesitate.

This puzzle has to do with what Jan Van Donk and I referred to as terminations, i.e., those sharp transitions which have carried the earth from maximum glacial conditions to maximum interglacial conditions on a time scale of 7 000 years or less (2). Perhaps they haunt me because as a graduate student I joined Bruce Heezen and Maurice Ewing in writing a paper entitled "Evidence for an abrupt change in climate close to 11 000 years ago" (3). I suspect that we, at that time, put our finger on a phenomenon which will bedevil theories of glaciation for many decades.

Were these terminations confined to the northern hemisphere they would not pose a problem. Not only did they also occur in the southern hemmisphere, but they seem to have been nearly synchronous with those in the northern hemisphere. Furthermore those in the south do not seem to have been driven by the demise of the vast glaciers of the north for there is good evidence that the southern hemisphere termination slightly preceded the melting of the great northern hemisphere ice sheets. If so, the planetary albedo and sea level changes associated with the disappearance of these northern ice sheets cannot be called upon to produce the southern hemisphere climate change. Rather, whatever event caused the warming which eventually led to the demise of the northern ice sheets struck both hemispheres at very nearly the same time. Lacking the large masses of excess glacial ice, the southern hemisphere appears to have completed its response to this forcing more promptly than did the northern hemisphere.

This apparent synchroneity poses a problem with regard to the claim that orbital forcing produces the 100 000 year periodicity in climate as well as the 21 000 and 41 000 year periodicities. As terminations appear to be a major aspect of the 100 000 cycle, any explanation for the 100 000 year cycle must also explain terminations. The problem is that, as shown in Figure 1, the seasonality history for the southern hemisphere is quite different from that for the northern hemisphere. This of course is the result of the fact that the two hemispheres are out of phase with regard to the precession cycle (~21 000 years) and in phase with regard to the tilt cycle (~41 000 years). In particular the northern hemisphere interval of strong seasonal contrast centered at 11 000 years ago is entirely absent in the southern hemisphere. This peak plays a key role in attempts such as those by Calder (4) and by Imbrie and Imbrie (5) to generate a climate record resembling the northern hemisphere ice volume curve from insolation curves. Without it there would be no way for orbital changes to trigger the sharp demise in ice volume close to 11 000 years. Thus I maintain that if the terminations are produced by insolation cycles they must reflect the impact of the seasonality history of the northern hemisphere.

If the northern hemisphere's message is not transmitted to the southern hemisphere though an ice albedo or sea level then how was it transmitted ? Any mechanism that involves the transfer of heat by the atmosphere from one hemisphere to the other runs smack into strong objections from climate models. GCM models show that the climate in the temperate and high latitude regions of each hemisphere is dependent almost entirely on the radiation budget of that hemisphere (Manabe pers.comm.). There is little impact of changes in the radiation budget of one hemisphere on the high latitude climate of the other hemisphere. This is not surprising since we know from tracer studies that the mixing time of air between the hemispheres is about 1 year, while the residence time of energy in the atmosphere is measured in weeks. Energy supplied by the sun will largely escape from the atmosphere before it moves from one hemisphere to the other. Thus not only must the ice albedo explanation be chucked but also any other explanation involving atmospheric transport between hemispheres.

Three possibilities remain. The first is that the heat transport of interest occurs in the sea (i.e., water heated in one hemisphere is cooled in the other). Turning on and off the North Atlantic Deep Water (NADW) is of course the prime candidate. Another possibility involves the climate-chemistry connection. The discovery that the CO_2 content of the atmosphere may have risen from 200 to 300 x 10^{-6} atm at the close of glacial time opens up a new vista (6, 7, 8). Regardless of the geographic location for the cause of this rise its effects would be global. If indeed the CO_2 content of the air did abruptly rise at the close of glacial time then its impact might have caused both the northern and southern hemisphere ice to retreat. The tough question remains. How did cycles in seasonality of the

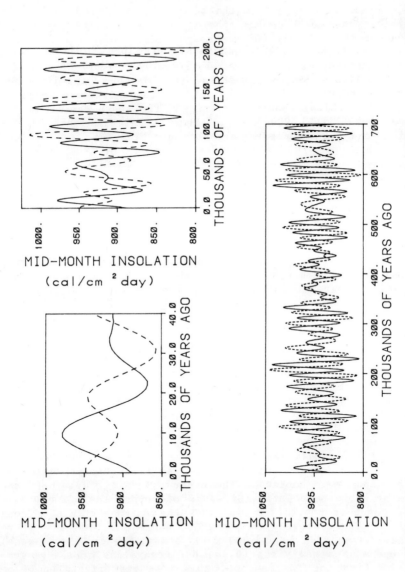

Figure 1 July insolation curves for 65° latitude in the northern (solid line) and southern (dashed line) hemispheres (20).

northern hemisphere lead to abrupt CO_2 changes ? The third possibility is that the terminations are not driven by insolation cycles but rather by some other phenomena, like abrupt solar output changes, which would affect the two hemispheres equally.

INTERHEMISPHERIC CORRELATIONS

Before trying to sort out these three possibilities it is important to summarize the evidence for the synchroneity of terminations in the two hemispheres. How firm is it ? There are three sources of evidence,

1) comparison of the ice extent records for South America with those for North America,
2) comparison of the faunal and floral records from deep sea cores raised from the northern Atlantic with those raised from the Antarctic,
3) comparison of the O^{18} records for borings in the Greenland and Devon Island ice sheets with those for borings in the Antarctic ice sheet.

In each of these three data sets there is indisputable evidence for a strong warming during the time interval 13 kyr to 8 kyr ago in both hemispheres. This almost everyone will accept. Can the timing be pinned down more accurately so that leads and lags be detected ? Can the relative magnitude of the climatic changes experienced in the two hemispheres be established ?

The record of the last major advance of the Laurentian ice sheet is well documented by hundreds of C^{14} ages (see, for example, 9). This ice sheet pushed into what is now Ohio about 25 000 years ago, reached a maximum about 20 000 years ago, began an oscillatory retreat about 15 000 years ago, was in full flight by 11 000 years ago and was largely gone by 8 000 years ago. Porter (10) and Mercer (11), working on the western side of the southern Andes, have gathered firm radiocarbon data for organic matter overridden by ice and for organic matter deposited on top of deposits formed during the retreat of these glaciers which makes a very strong case for synchroneity between the last advance and early retreat of the western Andean glaciers and glaciers of the northern hemisphere. These authors show that the ice reached its maximum extent close to 20 000 years ago and that the ice was largely gone by 11 000 years ago. Rather than repeat his detailed arguments here the reader is referred to the Porter and Mercer papers on this subject.

The micropaleontologic record in deep sea cores from the Antarctic can be closely correlated with that for the northern Atlantic through mutual correlation with the O^{18} record for benthic forams. As the latter record is thought to reflect an ocean wide change in the O^{18}/O^{16} ratio in sea water caused by the growth and retreat of the northern hemisphere ice caps it should be nearly synchronous throughout the deep sea. Not only does the benthic O^{18} record permit precise correlation of ecologic events at the two ends of the globe but it also permits both to be correlated with the demise of the great northern hemisphere ice sheets. This evidence (as illustrated in Figure 2) strongly points to the conclusion that the faunal and floral changes in the surface waters of the Antarctic at the close of glacial time slightly predated the O^{18} change which marked the major melting of the continental ice sheets (12, 13). Indeed these changes may well have a chronology consistent with that for the demise of the mountain glaciers in both hemispheres. By contrast, as shown by Ruddiman and Mc Intyre (14), the faunal and floral changes in the northern Atlantic followed the retreat of the adjacent continental ice sheets. This situation is atrributed to the influence of the ice caps on the meteorology of the North Atlantic.

Although accurate radiocarbon dating has yet to be accomplished on CO_2 from ice, Dansgaard and his coworkers using ice couplet data make a strong case that the transition from full glacial to full interglacial O^{18} values in the Greenland ice cores must have occurred roughly 10 500 years ago (17). Thus, while this event has not as yet been radiometrically dated it seems to match the time when the ice sheets in Scandinavia underwent their post Younger Dryas retreat. The O^{18} records in both the Camp Century and Dye 3 cores have a feature which quite likely corresponds to the Allerod-Younger Dryas climatic oscillation which immediately precedes this rapid final retreat. As Oeschger points out in this volume, the same sequence has been found in C^{14} dated O^{18} records for $CaCO_3$ from Swiss lake sediments (15). It also correlates with the oscillation in polar front seen by Ruddiman and McIntyre (14) in their study of cores from the northern Atlantic.

The situation for Antarctica is less certain. Seasonal couplets have not yet been utilized. Nevertheless, age estimates based on current accumulation rates and ice flow models yield about 10 000 years for the time of the O^{18} transition. It stretches the imagination to believe that this event differed by more than a few thousand years in age from that of the ecologic changes in the surface waters of the adjacent ocean or for the ice retreat in the Andes. However, the dating is clearly inadequate to say whether the air temperature change which produced the O^{18} change in the Antarctic ice slightly led or lagged the other events of interest.

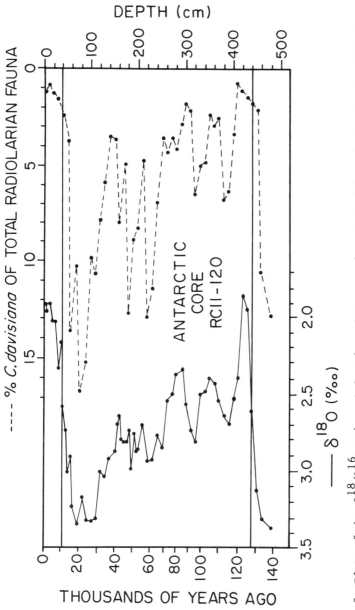

Figure 2 Plots of the O^{18}/O^{16} ratio and fraction of the species C. davisiana in the total radiolarian fauna as a function of depth in Antarctic core RC11-120. As can be seen the C. davisiana change leads the O^{18} change for both the termination which occured about 125 000 years ago and for that which occured about 11 000 years ago. The change in C. davisiana must reflect conditions in the overlying surface water while the O^{18} change is primarily the result of the melting of the northern hemisphere ice sheets.

Summarizing, then, all of these events appear to have occurred during a time of strong seasonal contrast in the northern hemisphere (i.e., associated with the 11 000 year ago summer radiation maximum). Those differences in timing which do show up to appear to be attributable to the sluggish behavior of the large Laurentian and Scandinavian ice masses. There is no evidence that the climatic change in southern hemisphere lagged that in northern hemisphere.

Only ice core evidence lends itself to contrasting the magnitude of the climate changes in the two hemispheres. As summarized in Figure 3, four of the ice core records yield roughly the same 0^{18} span between glacial and Holocene time (6-7 $°/\text{oo}$). Renaud and his colleagues in Grenoble (7,22) make a case (from air concentration data) that the Camp Century site underwent a marked decrease in elevation at the close of glacial time. If Renaud and his coworkers are correct then the glacial to interglacial air temperature change in the north polar region appears to have been comparable with that in the south polar region. If so, we are dealing with synchronous sudden climatic warmings of roughly the same magnitude !

POSSIBLE CAUSES FOR SOUTHERN HEMISPHERE TERMINAISONS

Of the three possible explanations given above let us consider the third. Do we have any evidence that might be used to exclude a solar origin for terminations ? We do. This evidence lies in the association of the terminations with northern hemisphere summer insolation maxima. As there is no way that the earth's insolation cycles could influence the sun, were fluctuations in solar output the cause of terminations then their distribution in time should be random with regard to the distribution in time of prominent summer insolation peaks. As can be seen in Figure 4 this does not seem to be the case. In this figure the SPECMAP composite 0^{18} curve for benthic forams (tuned in time in accord with techniques discussed in this volume) is shown together with the July insolation curve for 65 N. The first of each triad of strong summer insolation peaks is shown by an arrow ; these events are numbered with Roman numerals. As can be seen, four of the seven climatic cycles of the last 700 000 years have sharp terminations. These terminations are coincident in time with summer insolation peaks VII, IV, II and I. While for the other three cycles the situation is not so clear, in each case a prominent warming is seen in the 0^{18} record at the time of the appropriate insolation peak. This provides evidence that terminations are not solar in origin, and also that they are associated with peaks in northern hemisphere summer insolation.

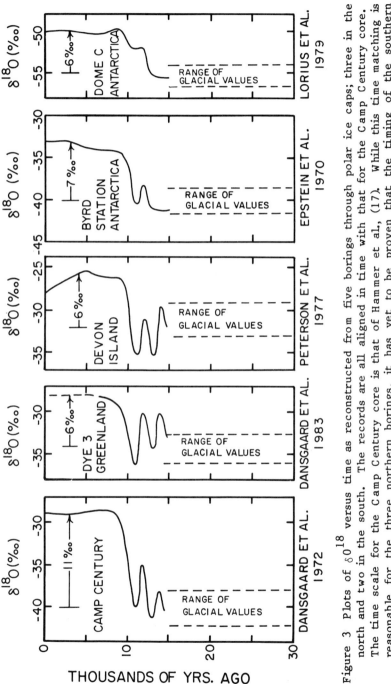

Figure 3 Plots of δO^{18} versus time as reconstructed from five borings through polar ice caps; three in the north and two in the south. The records are all aligned in time with that for the Camp Century core. The time scale for the Camp Century core is that of Hammer et al, (17). While this time matching is reasonable for the three northern borings, it has yet to be proven that the timing of the southern hemisphere change is the same as that for the northern hemisphere change.

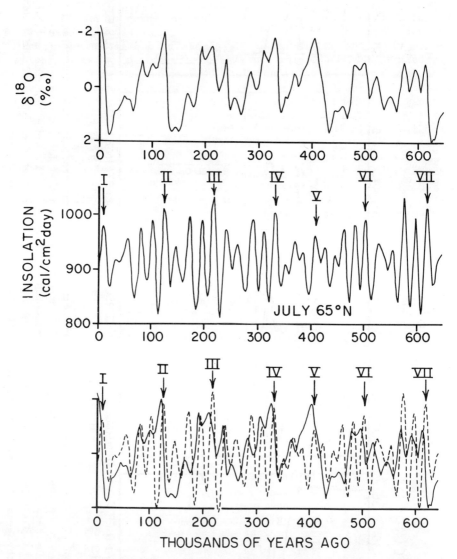

Figure 4 Relationship between the time history of O^{18} for benthic forams as averaged and tuned by the SPECMAP group (upper panel) and the summer insolation curve for 65° N (middle panel). The two curves are overlain in the lower panel. The arrows point to the first of each triad of strong summer insolation peaks. These triads coincide with maxima in the 100 000 year eccentricity cycle (20,21).

If the sun (and other extraterrestrial phenomena such as intergalactic dust) are eliminated in this way from contention then carbon dioxide becomes a very promising candidate. The problem is to conceive of a mechanism which might cause an abrupt change in concentration of this gas. If, as I have proposed (19), the CO_2 increase was caused by deposition of organic matter on shelves during the marine transgression accompanying the melting of the great northern hemisphere ice sheets, then it becomes difficult to explain the observation that the Antarctic faunal and floral changes led the O^{18} change in the sea. It also makes it difficult to explain the seeming sharpness of the termination in the southern hemisphere.

I have no solution to this mystery. Rather, I would like to emphasize the importance of learning more about the links between climate, ocean-atmosphere chemistry, and deep sea ventilation. This linkage is not only important to past climates but also in predicting the impacts of the coming CO_2 warming.

ACKNOWLEDGEMENTS

Over the last decade John Imbrie and I have indulged in many heated discussions about the origin of the 100 000 year cycle. The thoughts in this paper were triggered in this way. I thank John, not only for this intellectual stimulation, but also for providing the material for two of the figures (1 and 4) in this paper. I also appreciate the advice obtained from a number of my colleagues at Lamont (Hays, Fairbanks, McIntyre, Ruddiman, Morley, Kukla,...). It is also a pleasure not to have to thank any government agency for financial support. Thoughts are for free, no proposals need be written to get them rolling.

REFERENCES

1. Emiliani, C. : 1955, J. Geol., 63, pp. 538-578
2. Broecker, W.S. and Van Donk, J. : 1970, Rev. of Geophys. and Space Phys. 8, pp. 169-198.
3. Broecker, W.S., Ewing, M. and Heezen, B.C. : 1960, Amer. Jour. of Science 258, pp. 429-448.
4. Calder, N. : 1954, Nature, 252, pp. 216-218.
5. Imbrie, J. and Imbrie, J.Z. : 1980, Science, 207, pp. 943-953.
6. Berner, W., Stauffer, B. and Oeschger, H. : 1979, Nature, 275, pp. 53-55.
7. Delmas, R.J., Ascencio, J.-M. and Legrand, M. : 1980, Nature, 284, pp. 155-157.
8. Neftel, A., Oeschger, H., Swander, J., Stauffer, B. and Zumbrunn, R. : 1982, Nature, 295, pp. 220-223.

9. Dreimanus, A. : 1977, North America Annals of the New York Academy of Science, 228, pp. 70-89.
10. Porter, S.C. : 1981, Quaternary Research, 16, pp. 263-292.
11. Mercer : 1984, in "Climate Process and Climate Sensitivity", AGU, January 1984.
12. Hays, J.D., Lozano, J.A., Shackleton, N.J. and Irving, G. : 1976, in : "Investigation of Late Quaternary Paleooceanography and Paleoclimatology", R.M. Cline and J.D. Hays, (Eds.), Geol. Soc. Am. Mem. 145, pp. 337-372.
13. Hays, J.D., Imbrie J. and Shackleton, N.J. : 1976, Science, 194, pp. 1121-1132.
14. Ruddiman and Mc Intyre : 1981, Paleogeogr., -climatol. and -ecol., 35, pp. 145-214.
15. Eicher, V. and Siegenthaler, U. : 1976, Boreas, 5, pp. 109-117.
16. Dansgaard, W., Johnsen, S.J., Clausen, H.B., and Langway, C.C. : 1971, in "Late Cenozoic Glacial Ages", K.K. Turkian (Ed.), Yale University Press, pp. 37-56.
17. Hammer, C.U., Clausen, H.B., Dansgaard, W., Gundestrup, N., Johnsen, S.J., Reek, N. : 1978, J. Glaciology, 20, pp. 3-26.
18. Epstein, S., Sharp., R.P., and Gow, A.J. : 1970, Sciences, 168, p. 1570.
19. Broecker, W.S. : 1982, Geochim. Cosmochim. Acta, 46, pp. 1689-1705.
20. Berger, A. : 1978, J. Atmos. Sci., 35(12), pp. 2362-2367.
21. Berger, A. : 1978, Quaternary Research, 9, pp. 139-167.
22. Renaud, D. and Lebel, B. : 1979, Nature, vol. 281, no. 579, pp. 289-291.

[1]Contribution Number 3576 of Lamont-Doherty Geological Observatory/Columbia University.

MODELS FOR RECONSTRUCTING TEMPERATURE AND ICE VOLUME FROM OXYGEN ISOTOPE DATA

C. Covey and S.H. Schneider

National Center for Atmospheric Research, Boulder, CO 80307, USA.

Oxygen isotope ratios in ocean sediments are the primary evidence used to deduce the progression of climate changes during the Ice Ages (1). The major effect controlling the isotope ratios is believed to be preferential deposit of the lighter isotope, ^{16}O, onto ice sheets. It then follows that the $^{18}O/^{16}O$ ratio provides a measure of global ice volume. However, to reconstruct a time sequence of ice volume change, the preferential deposit of ^{16}O onto ice sheets must be estimated quantitatively and its flow through the ice sheets must be accounted for. With this goal in mind, we have constructed two models which can be used together to estimate the isotopic composition of the oceans, given a time series of global temperature and glacial ice volume. In this way--"climate" as input to the models, isotope "data" as output--we can evaluate various interpretations of the actual data in terms of climate changes.

Our models perform two separate tasks. An atmospheric model predicts the isotopic composition of snow, given the composition of the source water and the meteorological path between initial evaporation and final condensation. A glacial model follows the flow of isotopes through the ice sheets. The atmospheric model employs a "Lagrangian" approach in which the water vapor content and isotopic composition of a parcel of air are followed as the parcel moves along its prescribed thermodynamic path (Fig. 1). Results of the model depend strongly on (a) the degree to which nonequilibrium processes act in evaporation, (b) the height to which the parcel is raised after its formation, and (c) the temperature at which the parcel is precipitating. We have obtained empirical parameters for these processes by calibrating the model against isotope data from individual pre-

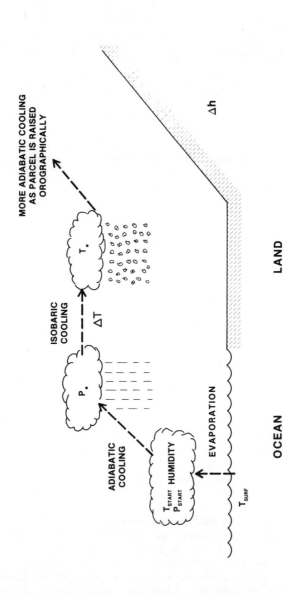

Figure 1 Main features of model for predicting isotopic composition of precipitation. A parcel of air and water vapor is formed by evaporation from the ocean at temperature T_{START}, raised adiabatically to pressure P_*, then cooled isobarically by amount ΔT. Upon encountering a continental ice sheet, the parcel is raised a distance equal to the height of the ice sheet at the point of precipitation (Δh). The $^{18}O/^{16}O$ ratios in the parcel and in precipitation from the parcel are followed throughout this process.

sent-day winter storms in New York State (2). We use air parcel trajectories computed by Philip Haagenson of NCAR from meteorological data (3). Parameters for processes (b) and (c) are obtained from the trajectories, while a parameter for process (a) is adjusted to fit the isotope observations (4).

Having determined in this way the necessary empirical parameters, we can extend our calculations to predict the isotopic composition of precipitation onto the Laurentide ice Sheet during the last glacial maximum. Our results for the "standard" case (Fig. 2) imply that near the ice divide, the depletion of ^{18}O based on the trajectory calculations relative to ocean water should have been of the order of 30 per mil. This result implies that the isotopic fractionation between ocean water and glacial ice is large enough so that most of the isotope variation seen in ocean sediments is due to changes in glacial ice volume--i.e., we verify the standard zeroth-order interpretation of the sediment data (5,6).

Nevertheless, the large sensitivity of isotope composition to the meteorological path followed by a parcel of water vapor introduces non-trivial uncertainties into the interpretation of isotope data. For example, ice deposited in the interior of Greenland was more depleted in ^{18}O during the last glacial maximum than it is today, by about 10 per mil (7). Our model indicates that this shift could be due to a 5K lowering of temperatures, a result consistent with the standard interpretation of the data. However, a plausible alternative explanation would be a shift in precipitation patterns leading to a greater (by 10 kPa) uplift in precipitating air masses (Fig. 2).

An important facet in the interpretation of isotope data from ocean sediments has been the measurement of isotope composition as a function of geological age. This data has in general been directly interpreted as a time series of glacial ice volume, but the long time that it takes isotopes to flow through a continental ice sheet may introduce distortions into the record. In order to study this possibility, we have constructed an ice flow model which employs an axisymmetric ice sheet centered on the North or South Pole. Use of a Glen-type flow law allows calculation of the size and shape of the ice sheet and flow velocities as functions of time, given the net accumulation of water mass. Thus the flow of isotopes can be traced over time once their deposition onto the ice is specified by the atmospheric model.

Prelminary results from the ice sheet model indicate that there are significant lag times between changes of ice volume and adjustment of the isotopic composition of the ice to its

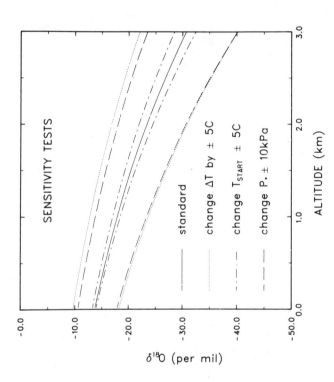

Figure 2 Isotopic composition of precipitation onto a continental ice sheet as a function of the height of the ice. Results are shown for seven model runs. "Standard" case uses values based on trajectory calculations for contemporary storms: $T_{START} = 10°C$, $P_* = 75$ kPa, $\Delta T = 5°C$, and an 8 per mil enhancement of isotopic fractionation due to kinetic effects during the initial evaporation. ^{18}O content of precipitation is decreased by either lowering the temperature or by raising the height to which the parcel is raised. The lowest two curves show that raising the parcel further (lowering P_* by 10 kPa) can have the same effect as lowering the surface temperature at which precipitation occurs (increasing ΔT by 5°C).

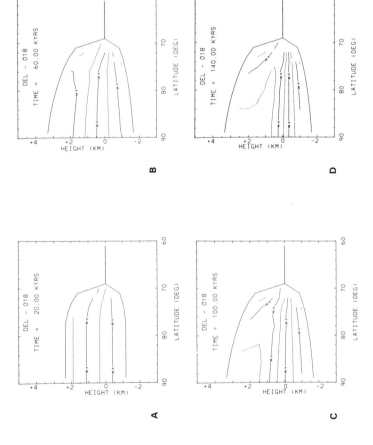

Figure 3 Model simulation of Antarctic ice sheet growth from an initial state of zero ice. The flow model is a simplified version of the model of Ledley (10). Empirical fits to today's observed values were used both for the net mass budget (11) and for the dependence of isotopic composition of snow on altitude (12). The results of this test run show that, while the ice sheet has nearly attained its equilibrium size by 60 kyr (b), the distribution of isotopic composition is still changing at 140 kyr (d).

ideal steady-state value. For example, in a simulation of the growth of the Antarctic ice sheet from zero initial ice (Fig. 3) the ice sheet attains its equilibrium size in about 60 kyr while isotopic composition has still not reached equilibrium after 140 kyr. The basic conclusion about uncertainties in lag times agrees with Mix and Ruddiman (8,9) and implies, as these authors point out, that the isotopic record in the oceanic sediments probably lags the actual changes in ice volume. Our results suggest that the lag times may be even more significant than Mix and Ruddiman propose since residence time of ice varies greatly with location in the ice sheet, with certain portions of the ice having a much longer residence time than the average period. For example, -20 per mil ice is trapped below sea level by isostatic sinking in our model (Fig. 3).

To summarize, we have constructed "reverse climate models" which predict the isotope proxy signal to be expected from given climatic conditions. The results from the models indicate that (1) ice core data reflects a combination of both surface temperature and uplift of precipitating air masses, with the relative contribution of each effect uncertain, and (2) oxygen isotopes in ocean sediments follow ice volume to "zeroth order", but significant lags (thousands of years) may well exist between ice volume changes and associated isotopic changes in the fossil record.

ACKNOWLEDGEMENTS

The authors thank Philip Haagenson for performing trajctory calculations, and William F. Ruddiman and Alan Mix for helpful comments on the manuscript.

REFERENCES

1. Hays, J.D., Imbrie, J., and Shackleton, N.J. : 1976, Science 194, pp. 1121-1132.
2. Gedzelman, S.D., and Lawrence, J.R. : 1982, J. Appl. Met. 21, pp. 1385-1404.
3. Haagenson, P.L., Shapiro, M.A., Middleton, P., and Laird, A.R. : 1981, J. Geophys. Res. 86, pp. 5231-5237.
4. Covey, C., and Haagenson, P.L. : 1983, submitted to J. Geophys. Res.
5. Shackleton, N.J. : 1967, Nature 215, pp. 15-17.
6. Dansgaard, W., and Tauber, H. : 1969, Science 166, pp. 499-502.
7. Dansgaard, W., Johnsen, S.J., Clausen, H.B., and Langway, C.C.Jr. : 1973, in : "The Late Cenozoic Glacial Ages", K.K. Turekian (Ed.), Yale University Press, New Haven.

8. Mix, A.C., and Ruddiman, W.F. : 1982, Quaternary Research, in press.
9. Mix, A.C., and Ruddiman, W.F. : 1984, in : "Milankovitch and Climate", A. Berger, J. Imbrie, J. Hays, G. Kukla, B. Saltzman (Eds), Reidel Publ. Company, Holland. This volume.
10. Ledley, T.S. : 1984, in : "Milankovitch and Climate", A. Berger, J. Imbrie, J. Hays, G. Kukla, B. Saltzman (Eds), Reidel Publ. Company, Holland. This volume, p. 581
11. Oerlemans, J. : 1982, Nature 297, pp. 550-553.
12. Morgan, V.I. : 1982, J. Glaciology 28, pp. 315-323.

INSOLATION GRADIENTS AND THE PALEOCLIMATIC RECORD

M.A. Young and R.S. Bradley

Department of Geology and Geography
University of Massachusetts
Amherst, MA 01003, USA

ABSTRACT

Hemispheric insolation gradients play an important role in driving the global atmospheric circulation, and may have contributed to the growth and decline of continental ice sheets by modulating the transport of moisture to high latitudes. Mid-monthly insolation differences between 30° and 90° latitude in each hemisphere were computed at 1000 years intervals for the past 150 000 years. Times of rapid ice build-up correpond to a distinctive seasonal pattern of insolation gradient deviations, with generally high gradients throughout the year, and follow closely times of strong autumn insolation gradients. The opposite patterns are observed at times of ice wastage.

INTRODUCTION

Throughout the development of the Milankovitch theory of climatic change, attention has focused on the insolation at critical latitudes and seasons which were thought to be particularly sensitive to glacierization and eventual ice sheet growth. In this paper, we will emphasize the more global view of orbitally-induced radiation variations that can be obtained by considering insolation gradients over substantial bands of latitudes.

Both insolation gradients and the boundary conditions on the climatic system (e.g. the presence or absence of ice sheets) influence hemispheric temperature gradients, which themselves are a major factor in determining the intensity of the extra-tropical atmospheric circulation. Several authors (e.g. 1,2) have examined these relationships and there is a consensus that stronger insolation gradients result in the displacement of the sub-tropical highs toward the equator, a more intense circumpolar westerly flow, and increased moisture transport to high latitudes. In contrast, weaker insolation gradients have the opposite effects, with a reduction in the moisture flux to artic and sub-artic regions. Berger (3) has suggested that the seasonal variation of the insolation gradient between 10 and 60°N, with its related poleward atmospheric transport of sensible and latent heat, provides a signature for the initiation of glacial and interglacial stages. Kutzbach, et al. (4) have also argued that the magnitude of astronomically-driven changes in insolation gradients is great enough to have produced significant climatic responses. Here, we make some simple time-domain comparisons between insolation gradient variations and the paleoclimatic record.

CIRCULATIONS, RESULTS, AND PALEOCLIMATIC COMPARISONS

The difference between mid-monthly insolation values at 30° and 90° latitude (hereafter referred to as the 30-90° insolation gradient) were computed for each hemisphere, at 1000 year intervals, over the past 150 kyr using the algorithm of Berger (5). Values are reported in units of W/m^2 (1 W/m^2 = 2.06 ly/day). Very similar results, qualitatively, were obtained for the 20-70° and 30-60° latitude bands in the northern hemisphere. Because the mid-monthly insolation values are based on constant true solar longitudes (0=March, 30=April, etc.), the length of time between successive monthly positions varies somewhat as the location of perihelion slides along the orbit. Despite this, we prefer this option over the calendar date option which artificially accomodates most of variation in the lengths of the astronomical seasons during the months around the autumnal equinox.

Contour plots of the deviations of the monthly gradient values from the appropriate monthly 150 kyr means are presented in Figure 1. A strong $\sim$23 kyr periodicity dominates the winter insolation gradients in each hemisphere, while the summer gradients exhibit fairly pure, though smaller amplitude, 40 kyr signals. Unlike insolation deviations, which must sum to approximately zero when all latitudes and seasons are considered, it is clear from Figure 1 that there have been times in the past when hemispheric insolation gradients were generally high (or low) throughout the year. These times are broadly synchronous between the two hemispheres, and a correspondence exists between times with generally high (low) 30-90° insolation gradients throughout the year in the northern hemisphere and times of increasing (decreasing) ice volume - as indicated by the $\delta^{18}O$

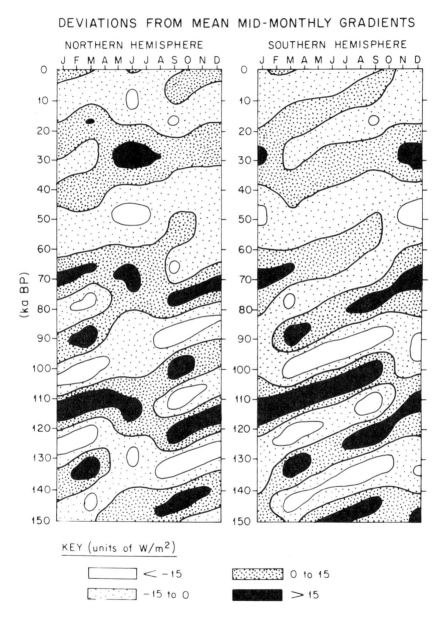

Figure 1 Deviations from monthly 150 kyr mean 30-90° insolation gradients for the Northern (left) and Southern (right) Hemipheres.

record of foraminiferal tests in marine sediments (6). The only dates with positive deviations in at least ten months occur at or near 23, 71, 116, and 141 kyr B.P.. Similarly widespread negative deviations occur at or near 11, 50, 84, and 128 kyr BP.

These relationships are more apparent in Figure 2, which shows the annual sums of the monthly gradient deviations as a function of time. Maxima and minima in the sums occur near the dates identified from Figure 1. Because of the uneven spacing of the mid-monthly values, and because the amplitude of variation of the gradients about the mean is largest during those months when the gradients are strongest, the sums of Figure 2 are weighted sums, with emphasis on the perihelion side of the orbit and on the equinoxes. However, we have produced very similar curves by using the calendar date option and by weighting the mid-monthly values by the amount of time spent in the vicinity of each position.

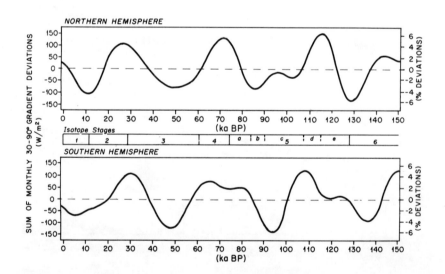

Figure 2 Annual sums of monthly 30-90° insolation gradient deviaions for the Northern (top) and Southern (bottom) Hemispheres. The oxygen isotope stage boundaries of (6) are shown for comparison.

DISCUSSION

A direct correspondence between insolation gradients and the ice volume record is unlikely to reflect accurately a true causal relationship since ice volume changes must certainly lag behind orbitally-induced radiation variations by several thousand years (7). If we assume a lag of about 5000 years, a more sensible correlation is apparent in which interglacials are related to periods of below average integrated annual insolation gradients, and glacials corespond with times of stronger than normal insolation gradients. This lagged correlation is in agreement with the work of Hays, et al. (8) who considered solar radiation receipts at 55-60°N to be the appropriate index of the orbital forcing function. However, energy balance models have often demonstrated that glacierization at these latitudes is not explicable solely in terms of in situ insolation changes (e.g. 9). A more powerful explanation involves both variations in hemispheric insolation gradients and subarctic insolation regimes.

To see how this might work, consider the seasonal pattern of insolation gradient deviations (Figure 3). We note that periods of rapid ice growth (e.g. 23 kyr B.P.) are marked by generally high gradients with maxima in the gradient deviations at the solstices, and follow times of strong autumn insolation gradients by several thousand years (Fig. 1). The correponding insolation regime is one of low summer and high winter receipts.

In contrast, times of rapid ice decay (e.g. 11 kyr B.P.) are distinguished by low gradients, with the most negative deviations at the solstices, and are preceded by times of weak autumn insolation gradients. These are times of high summer and low winter insolation.

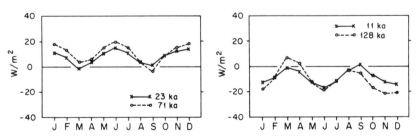

Figure 3 Seasonal patterns of deviations from monthly 150 kyr mean 30-90° insolation gradients at times of selected maxima (left) and minima (right) in the annual sums of figure 2.

If we associate insolation gradients directly with moisture transport to the subarctic, and consider that autumn precipitation is likely to play a particularly important role in glacierization (10), then a coherent picture emerges. Glacial maxima are thus seen to result from increased autumn insolation gradients (higher accumulation, peaking about 10 kyr before the glacial maximum) followed closely by periods of continued above average moisture transport and cool summers. Conversely, interglacials follow decreased autumn insolation gradients (lower accumulation) and subsequent periods of continuous low moisture transport coupled with warm summers (cf. 11). Positive feedback mechanisms within the climate system are also likely to be of considerable importance. For example, ice sheet growth would tend to strengthen hemispheric temperature gradients, reinforcing the insolation gradient changes.

CONCLUDING REMARKS

We have attempted to demonstrate the significance of changes in insolation gradients on climate during the past 150 kyr. Several difficulties remain in establishing such a relationship conclusively. First, the response of the climate system to insolation gradient variations must be more fully understood. Second, the effects of major changes in boundary conditions must be eliminated or corrected for. In this regard, the effects of insolation gradient changes are probably most accurately reflected in the paleoclimatic record when snow and ice cover are minimal, suggesting that the strong shift from very low to very high values of integrated annual insolation gradient, which occured between 125 kyr BP (when boundary conditions were probably similar to today) and 115 kyr BP, would be a good candidate for modelling experiments. Indeed, Royer et al. (12) have used a GCM to model the difference in climte between these dates. Varying only the insolation receipts, not the boundary conditions, they have found a significant zone of moisture surplus and lower temperatures in precisely those areas which are considered to have been important ice growth centers. Further studies along these lines should shed more light on the importance of insolation gradient changes.

REFERENCES

1. Lamb, H.H. : 1972, Climate, Present, Past and Future, vol. 1, Methuen, London.
2. Nicholson, S.E. and Flohn, H. : 1980, Climatic Change, 2, pp. 313-348.
3. Berger, A.L. : 1976, Long Term Variations of Daily and Monthly Insolation during the Last Ice Age. EOS, 57(4),

p. 254. Also Progress Report 1976/5, Institute of Astronomy and Geophysics, Catholic University of Louvain, Louvain-la-Neuve, Belgium.
4. Kutzbach, J.E., Bryson, R.E. and Shen, W.C. : 1968, Meteorological Monographs, 8, pp. 134-138.
5. Berger, A. : 1978, J. Atmos. Sci. 35(12), pp. 2362-2367. Also Contribution No 18, Institute of Astronomy and Geophysics, Catholic University of Louvain, Belgium.
6. Shackleton, N.J., and Opdyke, N.D. : 1973, Quaternary Research, 3, pp. 39-55.
7. Barry, R.G., Andrews, J.T. and Mahaffy, M.A. : 1975, Science, 190, pp. 979-981.
8. Hays, J.D., Imbrie, J. and Shackleton, N.J. : 1976, Science, 194, pp. 1121-1132.
9. Williams, L.D. : 1979, Arctic and Alpine Research, 11, pp. 443-456.
10. Kukla, G.J. : 1975, Nature, 253, pp. 600-603.
11. Ruddiman, W.F. and McIntyre, A. : 1981, Science, 212, pp. 617-627.
12. Royer, J.F., Deque, M. and Pestiaux, P. : 1984, in : "Milankovitch and Climate", A. Berger, J. Imbrie, J. Hays, G. Kukla, B. Saltzman (Eds), Reidel Publ. Company, Holland. This volume, p. 733

SUMMER TEMPERATURE VARIATIONS IN THE ARCTIC DURING INTERGLACIALS

R.G. Johnson

Honeywell Corporate Technology Center, Bloomington,
Minnesota, 55.420, USA

ABSTRACT

Modelling the effects of orbital variations on temperatures in the Canadian Arctic during interglacials like the present is relevant to the question of the time of onset of the next glaciation. Summer temperature profiles were calculated for Yellowknife (62.5°N-114°W) for the intervals from 125 000 years BP to 115 000 years BP and 6 000 years BP to 8 000 years AP. The model used numerically-integrated absorbed insolation values at 65°N, 45°N and 25°N. Albedos were calculated as a function of solar zenith angle with parameters consistent with satellite observations. At Yellowknife, not far from the Keewatin glacial nucleation area, model temperatures for July fall 4°C below present at 117 000 years BP near the Wisconsin/Weichselian glacial onset, and rise from a minimum near present to about 1.5°C above present at 8 000 years AP. This temperature rise suggests a continuation of the present major interglacial to at least 8 000 years AP.

INTRODUCTION

The Andrews-Mahaffy model of Laurentide ice-sheet initiation (1) indicates that a lowering of the glacial equilibrium line altitude of 400 m, consistent with a midsummer climate cooling of 4°C in the Canadian Arctic, could cause incomplete snow melting, and thus could initiate large-area ice-sheet growth such as occured following the last major interglacial that prevailed at 125 000 years BP. Deficits of summer insolation due to orbital changes may cause such cooling. To obtain the results of this paper, a model of the seasonal temperature profile in the Arctic was developed using a linear function of absorbed insolation, and fitted to today's temperatures at Yellowknife, N.W.T. Over the Yellowknife seasonal range of 45°C, this empirical temperature-generating function should be a good representation of the climate coupling between insolation inputs and the resulting temperatures. Small changes of insolation inputs due to orbital parameter differences at other ages should therefore yield good values of temperature profile changes. Thus, assuming that major atmospheric and oceanic circulation patterns are like those of today, as in the last 6 000 years, the model should give reliable indications of trends of the critical midsummer temperatures in the Yellowknife region.

PROCEDURES

A profile of daily absorbed insolation over the year at each latitude was calculated to provide temperature-generating function inputs over a 72 day interval prior to each point on the model temperature profile. The insolation points were calculated at four day intervals forward and backward from the summer solstice using orbital parameters of Berger (2). The value of the solar constant was 1.95 cal cm^{-2} min^{-1}. At each profile point, the rate of incident insolation was calculated at six degree hour angle intervals and each value multiplied by the absorbed fraction derived from the longitudinally averaged albedo. The sum over all intervals yielded the average daily absorbed insolation value for that profile point.

The absorbed fraction of solar radiation incident at the top of the atmosphere as a function of latitude was obtain using procedures discussed by Coakley (3). The total absorbed fraction, TAF, is given by:

TAF = (CLOUDF)[0.359 + 0.494 cos z − 0.258(1 − AFCLEARSKY)]

+ (1 − CLOUDF)(AFCLEARSKY) ,

where CLOUDF is the area fraction under clouds, z is the solar zenith angle, and where the absorbed fraction under a clear sky, AFCLEARSKY, is given by:

AFCLEARSKY = (OCEANF)(0.925 − 1.99x10^{-9} z$^{4.157}$)

+ (LANDF)(0.77 − 3.465x10^{-11} z$^{4.873}$)

+ (SN/ICEF)(0.48)

The expression for cloud-covered surface absorption of radiation was derived from the albedo expression given by Lian and Cess (4). The expression for the absorption by clear sky ocean area was derived from the zenith angle-dependent albedo of Coakley (3) and the clear sky reflectivities calculated by Braslau and Dave (5). The expression for the absorption by clear sky land area was derived from the calculations of Braslau and Dave assuming a land albedo of 0.25. The clear sky albedo for snow and ice was 0.52 with no zenith angle dependence. Cloud cover fractions were from London as listed by Cess (6), and ocean area fractions were largely from Sellers (7). Seasonal corrections were applied for variable snow and ice cover at 45°N and 65°N.

The temperature-generating function, A, relates the profile of absorbed radiation to the Yellowknife temperature profile. A covenient form of the function is:

$$A = T + 45° = (1/23.5)[R_1 + 0.5R_2 + 0.2R_3 + 0.1R_4 + F(R_5 + 0.4R_6 + 0.2R_7 + 0.3R_8 + 0.2R_9 + 0.1R_0)]$$

T is the daily maximum temperature in °C at a profile point, and R_1, R_2, R_3, and R_4 are values of absorbed daily radiation at 12, 32, 52, and 72 days respectively before the profile point at latitude 65°N. R_5, R_6, and R_7, and R_8, R_9, and R_0 are corresponding values at 45°N and 25°N latitude at 32, 52, and 72 days respectively. The 45°N and 25°N latitude radiation values were weighted by a function, F:

$$F = -0.01951 A + 1.3902 .$$

F represents the influence of meridional heat transport on high latitude radiation in winter relative to summer. The numerical terms and coefficients in these two equations were chosen to give values of T to match the present Yellowknife temperature profile (8) to within about one degree over the season. In the matching process, F and A satisfied both equations. For ages other than the present, present values of F were used, consistent with unchanged meridional temperature gradients under the assumption that the changes in lower latitude temperatures at other ages were similar to those at 65°N. Because such changes relative to the present are probably less at low latitudes than at high, the calculated temperature differences for the past ages may be slightly too large. Those calculated for the next 8 000 years may be too small, however, because of relatively stronger low latitude insolation then.

RESULTS

Figure 1 shows departures from the model of the present seasonal profile of temperature at Yellowknife for ages between 125 000 years BP and 115 000 years BP. Substantial temperature deficits during the July maximum temperature interval are shown by the model to begin at 120 000 years BP, and to reach a deficit of 4°C at 117 000 years BP. Such a temperature deficit, and less melting caused by less direct radiation absorption by ice and snow at Yellowknife and in the Keewatin and other glacial nucleation areas to the east, may well have prevented complete summer snow melting, and thus may have initiated a large area glacial build-up. This would be entirely consistent with the Andrews-Mahaffy model and with the marine record from which the initial Wisconsin/Weichselian ice-sheet built-up is inferred to have begun about 120 000 years BP (9).

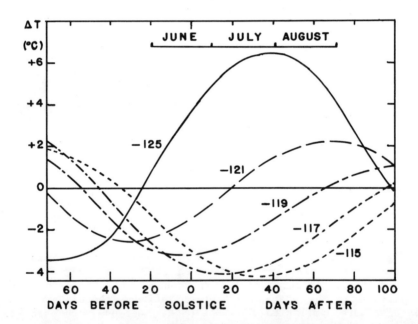

Figure 1 125 000 years BP to 115 000 years BP: Yellowknife model summer temperature differences relative to the model profile of present temperatures. The present July maximum of daily average maxima is 16°C.

Figure 2 shows the temperature departures for ages between 6 000 years BP and 8 000 years AP. The critical July temperature profile is presently near a minimum and shows a modest increase over the next 8 000 years, as does the July direct radiation absorption at 65°N. Thus, in contrast to the strong temperature deficits at Yellowknife that accompanied the initiation of the last major glaciation, the future temperature trends of the model do not suggest renewed ice-sheet growth, but rather a continuation of the present interglacial for at least 8 000 years.

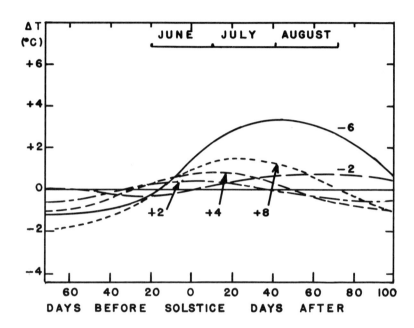

Figure 2   6 000 years BP to 8 000 years AP: Yellowknife model summer temperature differences relative to the model profile of present temperatures. Note the warming trend over future millenia.

REFERENCES

1. Andrews, J.T. and Mahaffy, M.A.W.: 1976, Quaternary Research 6, pp. 167-183.
2. Berger, A.L.: 1978, Quaternary Research 9, pp. 139-167.
3. Coakley, J.A.: 1979, J. Atmos. Sci. 36, pp. 260-269.
4. Lian, M.S. and Cess, R.D.: 1977, J. Atmos. Sci. 34, pp. 1058-1062.
5. Braslau, N. and Dave, J.V.: 1973, J. Appl. Meteor. 12, pp. 601-605.
6. Cess, R.D.: 1976, J. Atmos. Sci. 33: pp. 1831-1834.
7. Sellers, W.D.: 1965, in "Physical Climatology", University of Chicago Press, 272 pp.

8. Canadian Government Travel Bureau Data.
9. Chappell, J. and Thom, B.G.: 1978, Nature (London) 272, pp. 809-810.

CHANGES IN DECADALLY AVERAGED SEA SURFACE TEMPERATURE OVER THE WORLD 1861-1980

C. Folland and F. Kates

Meteorological Office, Bracknell, UK

ABSTRACT

Evidence is presented indicating that decadally averaged sea surface temperature (SST) has fluctuated significantly on the global space scale over the last century, with a minimum about 1911-20 and a maximum about 1951-60. The fluctuation has a range of about 0.6°C, though when averages are taken over the North Altantic Ocean alone it may exceed 0.8°C. These fluctuations on the time scale of a century seem large enough to be noted by climatologists concerned with the detailed mechanisms of ice-ages.

INTRODUCTION

The preliminary analyses presented here are based on a Meteorological Office archive of about 40×10^6 quality-controlled ship observations made since 1854 (The Meteorological Office Historical Sea Surface Temperature Data Set MOHSST (1)). The data are derived mainly from USA-11 SST tapes, the Meteorological Office collection of marine data and in recent years, telecommunicated data. Quality-controlled monthly averaged SST values have been estimated for 5° x 5° latitude/longitude tesserae ; a particular feature of the quality-control and estimation procedure is that it compensates for systematic changes in the positions of ships within each tessera. There is relatively little data south of 40°S so that MOHSST does not at present adequately represent the Southern Ocean.

ANALYSIS

SST in the 5° x 5° tesserae have been averaged separately over whole decades 1861-70, 1871-80, ... 1971-80 for each calendar month. These temperatures have been expressed as anomalies from the 1951-60 decadal average ; 1951-60 appears to be generally the warmest decade. The data in most areas is patchy in space and time so three separate analyses of the decadal anomalies have been carried out to test the sensitivity of the results to different ways of treating the data :

<u>Analysis I</u> : for each calendar month, a given 5° x 5° tessera is assigned a decadal SST anomaly only if at least three monthly SST averages exist in the 1951-60 reference period and three monthly SST averages exist in a given decade.

<u>Analysis II</u> : as Analysis I but at least seven monthly averages are required in 1951-60 and five in a given decade.

<u>Analysis III</u> : at least three monthly averages required in 1951-60 as in Analysis I but only one required in a given decade.

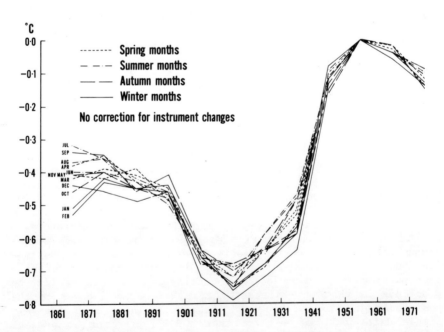

Figure 1 Quasi-global decadal anomaly of SST for each calendar month 1951-60 taken as base period. Analysis I.

A further problem with the data concerns the progressive change in measurement practice through the twentieth century from the almost exclusive use of sea-buckets early in the century to a substantial present use of engine-intake measurements. The change of practice would not matter except for evidence that on average engine-intake measurements give estimates of SST a few tenths of a degree warmer than the sea-bucket values e.g. (1) and (2). This is a difficult problem but its neglect can lead to artificial trends in SST ; provisional corrections are given in Section 3.

RESULTS

In Figures 1 and 2 near-globally averaged decadal temperature anomalies from 1861-70 to 1971-80 have been plotted from Analyses I and II respectively for each calendar month of the year. The results of Analysis III are similar and are not shown. The variation of near global anomaly with time is

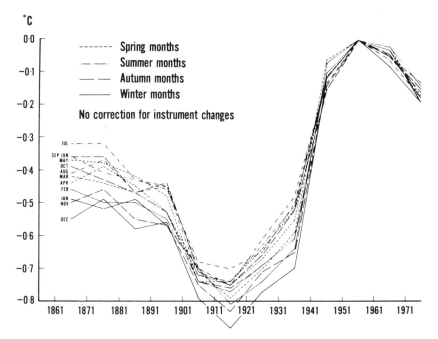

Figure 2 Quasi-global decadal anomaly of SST for each calendar month, 1951-60 taken as base period. Analysis II

surprisingly similar in each calendar month and similar in all three analyses. A minimum anomaly, mostly between -0.7°C and -0.8°C, is evident in 1911-20 while the 1951-60 decade is invariably the warmest. The late nineteenth century appears to be warmer than the early twentieth century and a modest cooling has taken place since 1951-60. Figure 3, derived from Analysis I only, shows the variation of decadally averaged SST taking all months together (solid lines) and also the results from applying provisional corrections for the artificial change in the SST resulting from instrumental changes (dotted - dashed line). The corrections are listed in Table 1 and arranged so that the correction for 1951-60 is defined to be zero. A feel for the magnitude of the corrections can be obtained as follows. The extensive investigation in Ref. (2) shows that the maximum artificial trend in SST likely to result from a change from the exclusive use of bucket measurements to that of engine-intake measurements averaged over all oceans is probably just over 0.3°C. It is believed that at the present time about one third of SST measurements are still made using buckets, so that an overall change in the corrections in Table 1 of -0.25°C is reasonable at this stage. The corrected curve has a reduced amplitude of about 0.6°C between the minimum, still in 1911-20,

Table 1 Variation of SST correction with time

	DECADE	CORRECTION °C
Up to	1921-30	+0.15
	1931-40	+0.10
	1941-50	+0.05
	1951-60	0.0
	1961-70	-0.05
	1971-80	-0.10

and the maximum now slightly more sharply defined in 1951-60. The curve is compared with the similar (perhaps slightly larger) amplitude of the Northern Hemisphere air temperature variations estimated by Jones, Wigley and Kelly (dotted line) (3) which includes restricted estimates of air temperature variations over the Nothern Hemisphere oceans. The SST maximum appears to lag that of the air temperature by about 15 years. Recently an analysis of decadal variations of a mix of land air temperature and SST representing much of the globe has been carried out by Paltridge and Woodruff (PW) (4). Their results

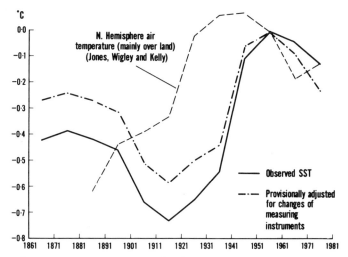

Figure 3 Inter-decadal variation of annual quasi-global SST. Base period 1951-60, Analysis I

also show a pronounced fluctuation of temperature with a minimum between 1900 and 1925 and a maximum between 1945 and 1970 but with an amplitude just under 1°C. PW's quasi-global temperature variation thus has similar phase to the SST variation found in this paper but with an amplitude (their Figure 5) about 50% larger. However, PW were only able to sample scattered areas of the globe.

Figure 4 shows a global map of 5° x 5° annually averaged SST anomalies for 1911-20 derived from Analysis I ; a uniform correction of +0.15°C has been applied (see Table 1). The map serves to show the geographical scope of Analysis I in 1911-20 and also gives an idea of the patterns of annually averaged global SST anomaly (from the 1951-60 values) near the early twentieth century minimum ; in particular it can be seen that nearly every tessera sampled is cooler than in 1951-60. (The annual anomalies shown in each tessera are an average of the monthly anomalies available). The North Atlantic was particularly cool in 1911-20 with negative anomalies everywhere and an average anomaly north of the equator of just over -0.8°C.

Figure 5 shows the results of a calculation of zonally averaged and corrected annual anomalies of SST for three selected decades 1881-90, 1911-20 and 1971-80. Each point represents a 5° latitude band and so the value plotted at 7.5° S represents the band 5°S to 10°S. 1911-20 is cooler in all latitudes than 1951-60 (the zero anomaly axis) while 1881-1890 is mostly warmer than 1911-20. Anomalies in 1971-80 show a more complex

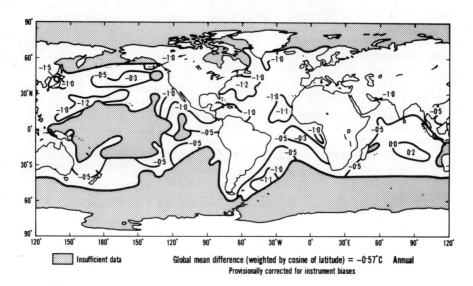

Figure 4 Mean decadal change in SST (1911-20)-(1951-60), Annual

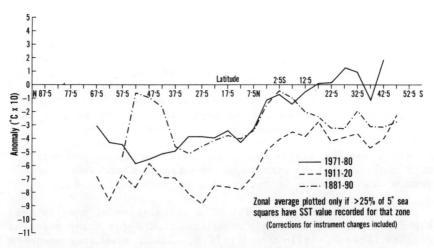

Figure 5 Annual zonally averaged anomaly (from 1951-60 normal) of SST. Analysis I

variation being generally below the 1951-60 average north of the equator but showing evidence of continued warming between 20°S and 40°S. A minimum long term variation of SST occuring near 10°S is confirmed by inspecting anomalies for the decades.

CONCLUSIONS

The variation of global SST over the last century appears to be of a magnitude and consistency that merits attention by climatologists interested in the mechanisms of the development and decay of Ice Ages. Evidence also exists of significant changes of Northern Hemisphere atmospheric circulation on a similar time-scale, notably over the North Atlantic where data are most complete (e.g. (5), (6)).

REFERENCES

1. Minhinick, J. : 1982, Sea Surface Temperature Data Sets in Met. O13. Unpublished, copy available in Meteorological Office Library.
2. James, R.W. and Fox, P.T. : 1972, Comparative Sea-Surface Temperature Measurements. WMO Rep. Marine Sci. Affairs, Rep. 5, WMO 336.
3. Jones, P.D., Wigley, T.M.L. and Kelly, P.M. : 1982, Mon. Wea. Rev. 110, pp.59-70.
4. Paltridge, G. and Woodruff, S. : 1981, Mon. Wea. Rev. 109, pp. 2427-2434.
5. Lamb, H.H. : 1972, Climate Present, Past, and Future. Vol. 1, Methuen.
6. Makrogiannis, T.J., Bloutsos, A.A. and Giles, B.D. : 1982, J. Clim. 2, pp. 159-169.

SEA-LEVEL FLUCTUATIONS AS EVIDENCE OF THE MILANKOVITCH CYCLES
AND OF THE PLANETARY-SOLAR MODULATION OF CLIMATE

R.W. Fairbridge

Department of Geological Sciences, Columbia University, New York 10027, USA

ABSTRACT

Eustatic fluctuations of sea level reflect Milankovitch cycles, but disclose also shorter periodicities, notably in the Holocene Epoch. Groups of beach ridges suggest possible solar-lunar cycles. A Hudson Bay storminess cycle of 45 yr varies also with longer-term periodicities that appear to match the 556-1112-1668 yr planetary and lunar cycles.

Terrestrial climate changes are believed to be triggered by three distinctive but interrelated exogenetic mechanisms, which are accelerated by endogenetic feedback processes. They are: (a) insolation modulated by terrestrial orbital motions, (b) solar radiation modulated by planetary torques and the Sun's circumbarycentric motions, and (c) geomagnetic screening of solar emanations modulated by Earth-Moon, "twin planet" motions. Evidence is essentially phenomenological and systematic study is called for. The Milankovitch insolation effects appear to be either amplified or retarded by cumulative climatic effects of the short-term cycles.

PART III

MODELLING LONG-TERM CLIMATIC VARIATIONS IN RESPONSE TO ASTRONOMICAL FORCING

SECTION 5 - GENERAL CIRCULATION MODELS

A SENSITIVITY EXPERIMENT TO ASTRONOMICAL FORCING WITH A SPECTRAL GCM : SIMULATION OF THE ANNUAL CYCLE AT 125 000 BP AND 115 000 BP

J.F. ROYER[1], M. DEQUE[1], P. PESTIAUX[2]

[1]Centre National de Recherches Météorologiques, F-31057 Toulouse Cedex, France

[2]UCL, Institut d'Astronomie et de Géophysique G. Lemaître, B-1348 Louvain-la-Neuve, Belgique

ABSTRACT

Our aim is to investigate, by means of a G.C.M., what can be the primary response of the atmosphere to the anomalies in the distribution of incident solar radiation by latitude and season introduced by the secular variations of the Earth's orbital elements (other boundary conditions being specified).

For such a sensitivity study we have chosen to simulate the annual cycle of climate at 125 000 years BP and 115 000 years BP, for which the insolation signatures computed at I.A.G. (Louvain-la-Neuve) are markedly different from the present one (perihelion in July and January with high eccentricity), whereas the surface boundary conditions can be reasonably assumed, similar to present ones (last interglacial).

The model used for this experiment is the 10-level sigma-coordinate primitive equations, global spectral model developped at E.E.R.M. (Paris) with a computational grid of 20x32 points. Its physical parameterizations include a simplified radiative scheme with diurnal variation of insolation, interactive cloudiness based on relative humidity, detailed hydrologic cycle and surface processes with computation of soil temperature, water content and snow cover.

We have compared two simulations of a complete annual cycle performed with the insolation parameters corresponding respectively to 125 kyr and to 115 kyr BP under identical seasonally varying boundary conditions (SST, albedo) specified according to actual climatology. Differences between the statistics of the two experiments show a characteristic response of the atmospheric temperatures and circulation to the insolation anomaly.

INTRODUCTION

The astronomical theory is probably the only theory of climatic change for which the variations of the external forcing can be known with great accuracy for long periods of the past covering several million years (1). Recent paleoclimatic results have apparently confirmed the idea that orbital variations by induced modifications of solar radiation distribution at the Earth's surface, have significantly influenced the Earth's Climate during the Pleistocene (2) and may be the cause of the glacial-interglacial transitions.

The basic problem that remains unsolved is to explain quantitatively the mechanisms by which such changes in the zonal and seasonal distribution of insolation can lead to the accumulation or ablation of the enormous quantities of ice stored in continental ice sheets. This problem seems of an extreme complexity and it clearly involves at least the 3 basic components of the climatic system : Atmosphere, Ocean and Cryosphere, which have very different time constants. The atmosphere is evidently a crucial link in this system since it is through its medium that the water is transported from the ocean surface, where it evaporates, to the ice sheets, where it will accumulate as snow and ice. To gain greater insight on its role in this system, one of the first steps is to study the impact on the atmosphere alone of the insolation changes produced by the orbital variations.

Among the best tools available at the present time to perform such sensitivity studies are General Circulation Models (GCM). Sensitivity experiments to changes of the solar constant have been performed for example by Wetherald and Manabe (3) and to changes of orbital parameters for 9 000 years BP by Kutzbach and Otto-Bliesner (4). In both studies cloudiness was specified by its present zonal values. In spite of this absence of feedback between cloudiness and the other dynamical variables of the model, they obtained a significant response showing the importance of the insolation variations on the atmospheric circulation.

In the present study we have chosen to perform two simulations of the general circulation of the atmosphere for a full annual cycle with the insolation conditions corresponding to 125 000 years BP and 115 000 years BP. These two experiments will be referred as 125 kyr and 115 kyr respectively.

The reason of the choice of these two particular periods is based on the following facts :

- They are the periods where are found the largest positive and negative deviations (from present conditions) of the July insolation in the Northern Hemisphere during the last 200 000 years, as can be seen on the figures published by Berger (5).

- They mark approximately the beginning and the end of isotopic substage 5e corresponding to the optimum of the last interglacial. 125 kyr corresponds to high sea-level stands (few meters higher than at present) recorded in marine terraces and coral reefs, whereas 115 kyr marks the beginning of glacial inception and return to colder conditions. During this whole period the North Atlantic polar front maintained a position far to the north-west like that of today (6). From faunal distribution of benthic foraminifera, Schnitker (7) has concluded that deep sea circulation during the last interglacial, 120 000 years ago, was similar to present conditions. Ruddiman and McIntyre (8) have given evidence that sea surface temperatures in the North Atlantic did not differ more than a few degrees from present day temperatures during this substage.

We have thus thought interesting to perform both simulations at 125 kyr and 115 kyr using the same pattern of sea-surface temperature and ice limits as observed to day. Even if subsequent more detailed paleoclimatic records show that this hypothesis is not realistic enough, our experiment will nevertheless provide an estimation of the basic sensitivity of the atmospheric system to the insolation variations alone, other boundary conditions being constant.

MODEL DESCRIPTION

The model used for this study is the spectral GCM developped at the EERM (Etablissement d'Etudes et de Recherches Météorologiques, Paris). It has been documented in details in a technical report (9). We shall give here only a short summary of its main characteristics.

The dynamical model

This model is based on the primitive equations of the atmosphere written in the sigma coordinate : $\sigma = p/p_s$ (p : pressure, p_s : surface pressure). The prognostic variables are the velocity streamfunction ψ and potential χ, temperature T and water vapour mixing ratio q for each of the 10 vertical levels, and the logarithm of surface pressure $X = \ln p_s$.

They are represented by a truncated expansion in surface spherical harmonics $Y_n^m = \exp(im\lambda) P_n^m(\mu)$ where λ : longitude, $\mu = \sin\phi$, ϕ : latitude, P_n^m : Legendre polynomials of the second kind, m : zonal wavenumber, n : spherical wavenumber.

Non-linear terms, including physical parameterizations, are computed on the latitude-longitude "Gaussian" grid associated with the chosen truncation. The transformation of the variables from real grid point values to complex spectral components are made efficiently by a Fast Fourier Transform (10).

The time-integration of the prognostic equations is performed in the spectral domain by the semi-implicit Leapfrog scheme described by Rochas et al. (11).

The general circulation version of this model makes use of a low resolution truncation of the "trapezoidal" type S 10-13 ($m \leq 10$, $n \leq 13$). The associated grid has 20 latitude circles (pseudo-equidistant in latitude $\Delta\phi=8.8-8.7$ between 83.3°N and S) with 32 points equidistant in longitude ($\Delta\lambda=11.25°$). The time step for the semi-implicit scheme is $\Delta t=60$ mn.

Some preliminary experiments with simplified physical parameterizations (Newtonian cooling with climatological zonal equilibrium temperatures) performed by Royer (12) have shown that a realistic simulation of the mean zonal fields could be obtained with this truncation.

The physical parameterizations

The present version of the model incorporates a complete representation of the main physical processes based on the parameterization scheme developped by Lepas et al. (13) for the New French Operationnal Ten-layer forecasting model (14).

Radiation transfer is parameterized as a function of water vapour using the "cooling to space" and "emissivity" approximations, which make the scheme fast enough for taking explicitly into account the diurnal variation of insolation.

Cloudiness is computed as a quadratic function of relative humidity for the 3 cloud layers (Nh : $0.2<\sigma<0.4$, Nm: $0.4<\sigma<0.7$, Nl: $0.7<\sigma<0.9$) and its effects on radiative fluxes is specified by empirical formulas.

The hydrologic cycle within the atmosphere is described by the following processes :

- synoptic precipitations by large scale saturation with evaporation of the falling rain in unsaturated layers,
- deep convection, by a modified Kuo's scheme,
- moist convective adjustment with critical lapse rate function of the relative humidity of the layers.

Horizontal diffusion is expressed in spectral space by a ∇^4 operator applied to the prognostic variables ψ, χ, T, q.

Surface processes include :

- a prognostic equation for soil temperature taking into account heat conduction into the ground,
- surface fluxes are computed by the bulk aerodynamics formulas,
- surface hydrology : soil water content, W.

Mass of fresh snow Wnj and old snow Wnv (including ice) are prognostic variables computed from balance equations. The snow metamorphism (Wnj $\rightarrow$ Wnv) is assumed to be proportional to the mass of fresh snow with a characteristic time of 4.5 days. Albedo and thermal properties of the ground depend through empirical relations of the predicted values of soil water content W and snow depths Wnj and Wnv.

BOUNDARY CONDITIONS

We have performed with the model described above 3 experiments with different insolation conditions and identical surface boundary conditions : a control experiment and 2 perturbed experiments corresponding to the 125 kyr and 115 kyr insolation. The control experiment (D4) is a 5 year simulation of the present climate that will be used to provide significance levels to the deviations obtained in the 125 kyr and 115 kyr runs.

Insolation parameters

Solar radiation is computed at each time step as a function of latitude and local time of each grid point from the values of solar declination δ and normalized solar intensity : $S' = S/S_o = (a/r)^2$. (S_o is the solar constant, r the earth-sun distance, a the semi-major axis of the elliptic orbit). The 2 solar parameters δ and S' are updated each day and computed by the classical formulas as a function of the 3 orbital elements of the earth : obliquity ε, eccentricity e and longitude of the perihelion $\tilde{\omega}$, and position of the earth on its orbit λ (longitude relative to vernal equinox). The values of the orbital parameters adopted in our 3 experiments are displayed in Table 1. They have been computed according to the formulas of Berger (15) by using a program that he kindly made available to us.

Table 1 Values of the earth's orbital elements in the 3 experiments for solar insolation computations.

	125 kyr	115 kyr	Present
Obliquity	23°.798	22°.405	23°.447
Eccentricity	0.04001	0.04142	0.01672
Longitude of the perihelion	307°.14	110°.9	102°.0

In all the annual cycles the position of the earth on its orbit is computed as a function of calendar date relative to a vernal equinox fixed at March 21st (15).

Monthly mean insolation anomalies relative to present for 125 kyr and 115 kyr are represented as a function of latitude and month on Figure 1 as indicated by Berger in (5). At 125 kyr we clearly see a large positive anomaly reaching more than 50 W/m^2 in July in middle and high latitudes of the Northern Hemisphere, extending in the Southern Hemisphere in September-October and a large negative anomaly in January in the Southern Hemisphere. At 115 kyr the situation is opposite with a negative anomaly in July and a positive one in January having about half the intensity of the 125 kyr anomaly. Such large anomaly results directly from larger eccentricity values and passing at perihelion around mid-July and mid-January respectively. If we compare the 115 kyr insolation conditions with the 125 kyr we find an excess reaching 16% in January and a similar deficit in July. These 2 maps can be viewed as the forcing functions of the 2 perturbed experiments relative to the control case.

Surface boundary conditions

As explained in the introduction all other boundary conditions were identically specified in the 3 runs. Orography and land-sea distribution were fixed. Sea surface temperature and sea ice extent and thickness were updated each day during the integration in order to take into account their seasonal variation. The climatological values (16) were supposed to apply to

SENSITIVITY EXPERIMENT TO ASTRONOMICAL FORCING WITH A SPECTRAL GCM 739

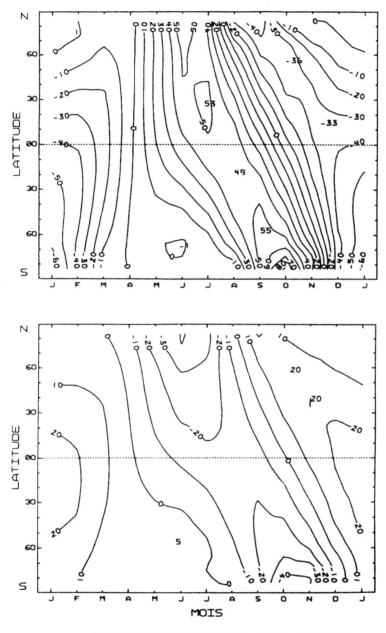

RAYONNEMENT SOLAIRE W/M2 115KBP - PRESENT

Figure 1 Distribution by month and latitude of the monthly mean insolation deviation from the present state (in W/m^2). A): 125 000 BP minus present; B): 115 000 BP minus present

the 15th of each month and a linear interpolation was used to define intermediate values.

The same procedure was used for surface albedo. The 125 kyr and 115 kyr experiments were started from the same initial situation taken as 1st December of the second year of the control run. The first month of their integration (December) was considered to be a transitory adaptation period of the model to the new insolation conditions and discarded from the subsequent statistical analysis which was based on a complete year simulation starting in January plus January of the following year.

RESULTS

We have analysed the results of the 2 simulations by computing the difference of the 115 kyr minus the 125 kyr monthly means. Such differences show the response of the Climate simulated by the model to the change in insolation conditions at the beginning and end of the last interglacial isotopic substage 5e which are of particular interest in the context of the Milankovitch theory.

Global and seasonal means over continents and oceans

In order to show clearly the contrast between continental regions, where surface temperature is computed by the model from an energy balance equation, and oceanic regions, where the sea surface temperature is specified, we have given in Table 2 spatial and temporal averages computed separately over continents and oceans.

The cryosphere is included in the definition of the continental areas (snow, permanent ice and ice sheets) and oceanic areas (ice pack). However, even though the ice pack extent and thickness are specified, its surface temperature is computed from an energy balance equation like in the continental case.

It is specially important to analyse the high latitudes in both hemispheres since it is where we find most of the cryosphere. We have thus considered 3 regions to perform spatial averaging :

- the Northern High Latitudes (NHL), represented by the zone northward of 53°N which corresponds to the 4 latitude circles closer to the pole in our model.
- the Middle and Low Latitudes (MIL), zone between 53°N and 53°S.
- the Southern High Latitudes (SHL) from 53°S to 90°S.

In the bottom row are included the global averages.

The annual variation has been subdivided in the 4 seasons to which we shall conventionnally refer below according to their northern hemisphere definition. We have represented them by 3 month averages following the usual practice in climatology : spring (MAM), summer (JJA), autumn (SON) and winter (DJF). The annual mean is shown in the last column. The values in Table 2 are the differences between the statistics of the 2 simulations (115 kyr minus 125 kyr).

Table 2 Difference between the 115 kyr and 125 kyr simulations averaged over latitude zones and seasons for continent and ocean.

NHL : Northern High Latitudes (poleward of 53°N)
MIL : Middle and Low Latitudes (53°N-53°S)
SHL : Southern High Latitudes
Global

MAM : March April May
JJA : June July August
SON : September October November
DJF : December January February
annual

A) CONTINENT

Sea-level pressure (mb)

	MAM	JJA	SON	DJF	ANNUAL
NHL	1.2	4.5	-3.4	-1.7	0.2
MIL	-0.7	2.5	-1.1	-1.1	-0.1
SHL	-3.9	-4.7	2.7	-8.6	-3.6
Global	-0.6	2.4	-1.3	-1.8	-0.3

Cloudiness (in%)

	MAM	JJA	SON	DJF	ANNUAL
NHL	1.5	8.1	2.4	2.2	3.6
MIL	0.5	-2.8	-1.6	0.9	-0.8
SHL	-1.7	-2.5	1.7	0.6	-0.5
Global	0.5	-0.6	-0.6	1.1	0.1

Solar radiation absorbed at the surface (W/m^2)

	MAM	JJA	SON	DJF	ANNUAL
NHL	-3.9	-34.2	12.1	0.7	-6.4

MIL	2.0	-25.0	11.1	22.1	2.4
SHL	0.5	-0.2	-7.9	4.6	-0.8
Global	0.7	-25.1	9.9	16.5	0.4

Temperature at 850 mb (°C)

	MAM	JJA	SON	DJF	ANNUAL
NHL	0.2	-4.8	1.1	0.5	-0.7
MIL	0.8	-2.9	0.9	2.0	0.2
SHL	-1.0	-0.2	-2.5	2.1	-0.4
Global	0.5	-3.1	0.7	1.7	-0.0

Saturation deficit T-Td at 850 mb (°C)

	MAM	JJA	SON	DJF	ANNUAL
NHL	0.1	-1.9	-0.9	-0.7	-0.8
MIL	0.4	-0.5	-0.2	-0.9	-0.3
SHL	0.3	-1.0	0.8	-0.2	-0.0
Global	0.4	-0.8	-0.3	-0.8	-0.4

B) OCEAN

Sea-level pressure (mb)

	MAM	JJA	SON	DJF	ANNUAL
NHL	0.3	1.5	-4.2	-1.7	-1.0
MIL	0.2	-1.0	0.3	0.7	0.1
SHL	-3.8	-2.2	0.5	-4.7	-2.5
Global	-0.2	-1.0	0.1	-0.0	-0.3

Cloudiness (in %)

	MAM	JJA	SON	DJF	ANNUAL
NHL	2.6	1.1	0.9	0.3	1.3
MIL	0.5	1.6	-1.2	-0.8	0.0
SHL	1.1	2.0	-1.0	0.6	0.7
Global	0.7	1.6	-1.0	-0.6	0.2

Solar radiation absorbed at the surface (W/m^2)

	MAM	JJA	SON	DJF	ANNUAL
NHL	-6.9	-20.7	13.1	1.4	-3.4
MIL	2.6	-30.8	-1.3	33.1	0.7
SHL	2.4	-4.1	-19.1	21.1	0.0
Global	2.0	-27.2	-2.5	29.8	0.4

Temperature at 850 mb (°C)

	MAM	JJA	SON	DJF	ANNUAL
NHL	0.4	-3.6	-0.5	0.3	-0.9
MIL	0.1	-0.7	-0.1	0.3	-0.1
SHL	-0.0	0.1	-1.4	0.6	-0.2
Global	0.1	-0.8	-0.3	0.3	-0.2

Saturation deficit T-Td at 850 mb (°C)

	MAM	JJA	SON	DJF	ANNUAL
NHL	0.1	-0.1	-0.7	-0.6	-0.3
MIL	-0.0	-1.2	-0.2	0.4	-0.3
SHL	-0.6	-1.3	1.4	0.1	-0.1
Global	0.1	-0.8	-0.3	0.3	-0.2

For the solar radiation absorbed at the surface, we see that in global mean the large deficit of 25 W/m^2 over continents in summer is counterbalanced by a surplus in autumn and winter giving in annual mean an unsignificantly slight increase (0.4 W/m^2). However, for the Northern high latitudes we find an annual mean deficit of 64 W/m^2 over continent (3.4 W/m^2 over the ocean), because the summer deficit (35 W/m^2 over continent and 21 W/m^2 over ocean) is only partially compensated by an excess in autumn (12 and 13 W/m^2 respectively), but not in the winter when the insolation is very small at such latitudes. We can notice that in global average the seasonal amplitude is slightly greater over the oceans. This fact can be explained by the different response of cloudiness over ocean and continent.

In Northern high latitudes we find a cloudiness increase having a maximum of 8% in summer over the continents, and a smaller increase with a spring maximum of 2.6% over the oceans. However, in the middle and low latitude zone the response is different: with the decrease of summer insolation we observe a decrease in cloudiness (2.8%) over the continents and a smaller increase (1.6%) over the oceans.

Nevertheless, over the Northern high latitudes, in response to the insolation deficit mentioned before, there is a decrease in annual mean of 0.7°C over continents and 0.9°C over ocean. This result can be interpreted in the frame of the Milankovitch theory by saying that the cold summers in the high latitudes of the Northern hemisphere resulting from an insolation deficit when the Earth is at aphelion in July are not compensated by warmer winters. The radiation balance of the high latitudes seems strongly affected by such changes in the annual distribution of solar radiation and the average temperature of this zone reacts consequently.

If we look at the saturation deficit expressed by the difference between the air temperature T and dew-point temperature Td, at the 850 mb level for example, we see a general decrease showing that the atmosphere is closer to saturation in the 115 kyr simulation. It may be interesting to relate this result with the findings of Jouzel et al. (17) who deduce from isotopic analysis of ice cores that the atmosphere had higher relative humidity values during the last glacial maximum.

Precipitations are not changed in annual mean. In response to the changes in monsoonal circulation there is a decrease of 0.9 mm/day in summer precipitation over MIL continents, a corresponding increase of 0.3 over oceans and opposite variations of smaller amplitude in winter. We notice a small positive increase (0.1 mm/day) in continental Northern high latitudes in autumn and winter which could fall as snow. The evaporation follows a pattern largely similar to that of precipitation. There is a reduction in summer in NHL continents (0.4 mm/day) due to colder surface temperatures. This small positive feedback over the oceans tends to amplify slightly the amplitude of the annual variation of solar radiation absorbed at the surface. Such opposite variations of cloudiness are the consequence of circulation changes that are clearly seen in sea-level pressure. In the MIL zone we have over continents an increase of 2.5 mb in summer and a decrease in the other season. Over the oceans the response is opposite and of half the amplitude, with a 1 mb decrease in summer.

The similarity of the seasonal behaviour over ocean and continent in the NHL zone can be attributed to the influence of the cryosphere in this region. The different response of sea-level pressure over ocean and continent at lower latitudes can be interpreted in term of reduction or intensification of monsoonal type circulation patterns, as in the simulation of Kutzbach and Otto-Bliesner (4). This contrast between oceanic and continental regions has its origin in their different temperature response to the change in the insolation conditions: the surface temperature on the ground (as well as on ice) can adjust rapidly to the new insolation whereas the sea surface temperature does not change since it is specified independently (which is in fact a simplified picture of the large thermal inertia of the ocean). At high levels the contrast in the response of the air temperature over continents and over oceans becomes attenuated but still remains apparent.

At the 850 mb level, for example, we see clearly that the air temperature responds directly to the insolation deviations with a decrease in summer of about 3°C over the continents and a smaller decrease of 0.8°C over the oceans. In the MIL zone this is compensated by a 2°C increase in winter over the continents, and in global annual average the temperature difference is nearly zero.

On the whole the hydric balance (precipitation minus evaporation) is not changed in global annual mean, but is slightly positive for NHL continent in summer (0.3 mm/day) and in winter

(0.1 mm/day) as a consequence of reduced evaporation and higher precipitation respectively. All such variations could contribute to the formation of an ice cover on the Northern hemisphere continental areas in high latitudes.

Annual variation of the zonally averaged anomalies over the continents at 125 kyr and 115 kyr

In order to have more details of the model response to the insolation forcing we have drawn several figures of the latitude-month distributions of the anomalies relative to the control case (present climate). With the 5 years of simulations in the control case we can estimate as well the standard deviation of the monthly zonal mean; this gives us the possibility of testing the statistical significance of the model response by means of a Student t test according to the method described by Chervin and Schneider (18). For our experiment the deviations from the 5 year mean control case are found significant at the 5% level when they exceed 3.045 standard deviations. On our figures we have shown with hatching the areas where the differences are significant at the 5% level for this statistical test.

The latitude-month distributions of the different fields show a variable degree of similarity with the distribution of the solar forcing (Fig. 1). The extent of the statistically significant areas is systematically larger in the 125 kyr than in the 115 kyr simulation, which is direct consequence of the larger solar forcing in the former simulation. The extent of such areas differs markedly according to the variables considered, in relation to their more or less direct physical link with the forcing.

The pattern of solar radiation absorbed at the surface of the continents (Fig. 2) is largely similar to the insolation forcing in middle and high latitudes. Some discrepancies that appear in the intertropical zone, such as the deficit in April, May and August near 15°N for 125 kyr, are the consequence of cloudiness changes (Fig. 6).

The temperature variations at 850 mb (Fig. 3) follow closely the radiation patterns. For 125 kyr we have a significant warming in summer from the pole to 20°N with maxima of about 4°C at 30°N and 60°N, and a cooling starting in October at 45°N and extending in winter and spring towards lower subtropical latitudes. At 115 kyr we have the opposite situation with a cooling in summer reaching 3°C around 60°N and a warming in winter and spring around 30°N. We see that near 60°N the colder conditions remain nearly the whole year with only a short but significant warming in October, whereas in the equatorial zone the temperature changes are very small.

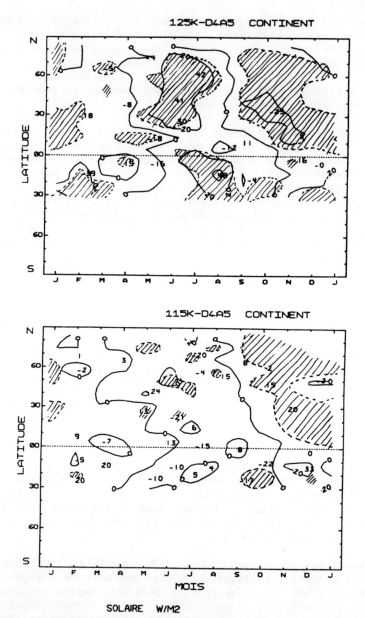

Figure 2 Simulated latitude-month distribution of the difference of the monthly mean solar radiation absorbed at the surface (in W/m^2) zonally averaged over the continents. Shading shows areas where the difference exceeds 3 standard deviations.
A): 125 kyr minus present; B): 115 kyr minus present

The sea-level pressure response (Fig. 4) is correlated directly with the temperature response only in summer where warming and cooling are associated respectively with a pressure decrease and increase which are related to monsoonal type circulation changes over the continents. The pressure response in other seasons is more complicated and not easily interpreted. In high latitudes the variations are large but rarely significant. For the latitude zone near 60°N we tend to have a pressure anomaly of the same sign at the beginning of winter as in summer, and an opposite anomaly in autumn and at the end of winter. For 115 kyr this gives more cyclonic situations in October and in February, and more anticyclonic conditions in summer and in December.

The amplitude of the saturation deficit T-Td at 850 mb (Fig. 5) shows for 125 kyr a tendency towards a dryer atmosphere in middle and high latitudes in summer, and in subtropical latitudes in winter. For 115 kyr the tendency is toward an atmosphere closer to saturation, except for the high latitudes in winter and spring and a zone in the Northern subtropics in May-June.

The changes in cloudiness (Fig. 6) cannot be related in any simple way to the solar forcing and show a complicated pattern, as well as the precipitations (Fig. 7). Both fields show small areas of significant variations mainly in the tropical zone. There is however a general tendency toward an increase of cloudiness in middle latitudes at 115 kyr with an increase of precipitation.

For many variables zonal averages over the oceans show similar patterns as over the continents, only with somewhat reduced amplitudes. However for some quantities like surface pressure, cloudiness and hydric balance (Fig. 8 to 10) have opposite reactions over continents and over oceans. For the sea-level pressure we clearly see in summer an increase over continents (maximum 7 mb) and a decrease over oceans (maximum 5 mb) in middle latitudes. In the high latitudes the variations are largely similar with more cyclonic conditions in autumn and at the end of winter. The same pattern is seen in cloudiness with a general increase in high latitudes. The hydric balance tends to be more positive over the continents in high latitudes in spring and summer.

The above figures give evidence that the response to the insolation variations is markedly different according to the latitude zone considered.

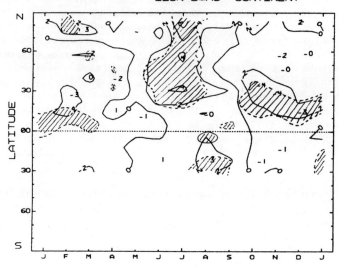

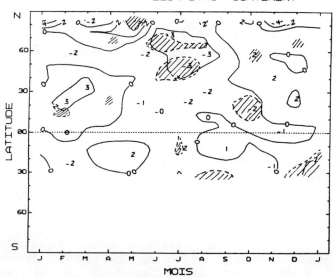

Figure 3 Simulated latitude-month distribution of the difference of monthly mean temperature at the 850 mbar level (in °C) zonally averaged over the continents. Shading shows the points where the difference exceeds 3 standard deviations.
A): 125 kyr minus present; B): 115 kyr minus present

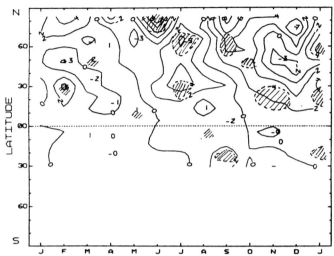

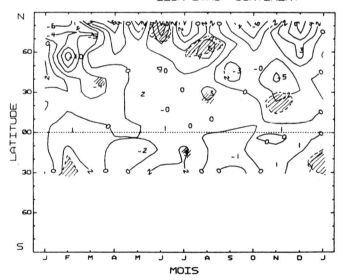

Figure 4   Simulated latitude-month distribution of the difference of monthly mean sea level pressure (in mbar, or hPa) zonally averaged over the continents. Shading shows the regions where the difference exceeds 3 standard deviations.
A): 125 kyr minus present;   B): 115 kyr minus present

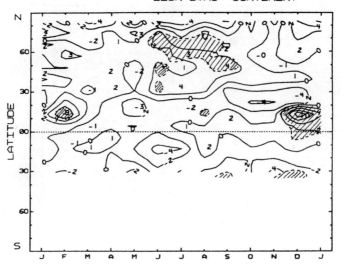

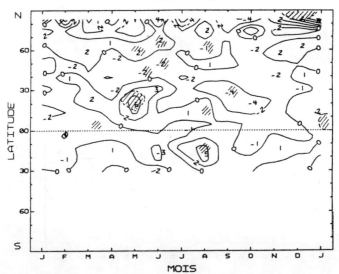

Figure 5 Simulated latitude-month distribution of the difference of monthly mean saturation deficit T-Td (in °C) at 850 mb zonally averaged over the continents. Shading shows the regions where the difference exceeds 3 standard deviations.
A): 125 kyr minus present;   B): 115 kyr minus present

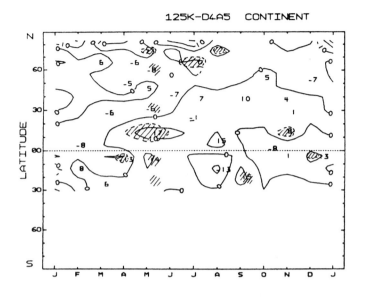

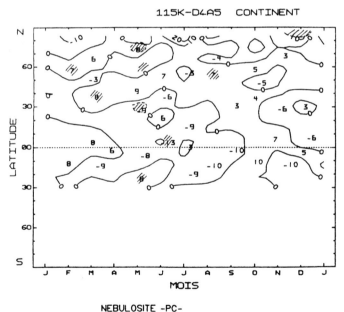

NEBULOSITE -PC-

Figure 6 Simulated latitude-month distribution of the difference of monthly mean cloudiness (in %) zonally averaged over the continents. Shading shows areas where the difference exceeds 3 standard deviations.
A): 125 kyr minus present; B): 115 kyr minus present

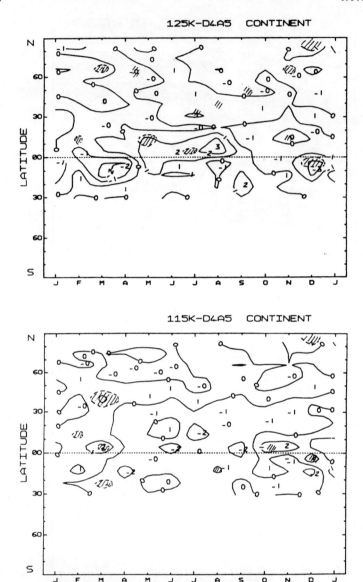

Figure 7   Simulated latitude-month distribution of the difference of monthly mean precipitation (in mm/day) zonally averaged over the continents. Shading shows areas where the difference exceeds 3 standard deviations.
A): 125 kyr minus present; B): 115 kyr minus present

SENSITIVITY EXPERIMENT TO ASTRONOMICAL FORCING WITH A SPECTRAL GCM 753

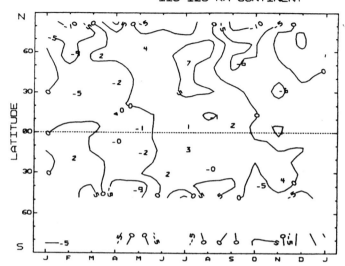

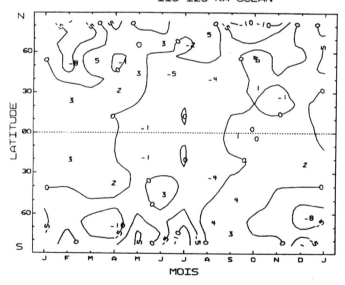

Figure 8 Latitude month distribution of the difference between the 115 kyr and 125 kyr simulations of monthly mean sea level pressure (in mbar, or hPa).
A): zonally averaged over the continents; B): zonally averaged over the oceans

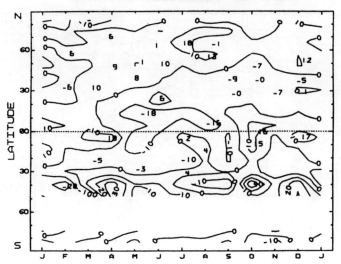

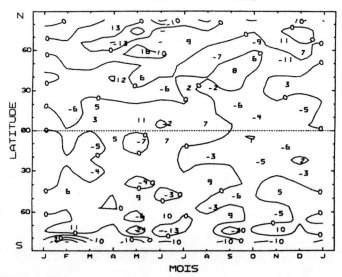

Figure 9   Latitude month distribution of the difference (115 kyr minus 125 kyr) of monthly mean cloudiness (in %).
A): zonally averaged over the continents; B): zonally averaged over the oceans

115-125 KA CONTINENT

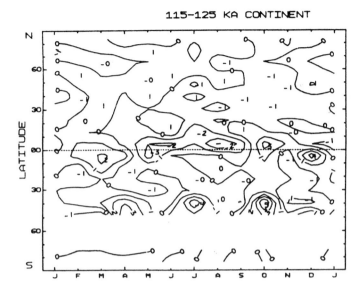

115-125 KA OCEAN

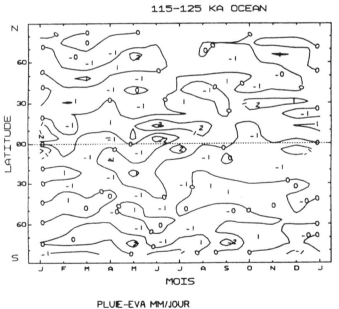

PLUIE-EVA MM/JOUR

Figure 10 Latitude month distribution of the difference (115 kyr minus 125 kyr) of monthly mean hydric balance at the surface (precipitation minus evaporation in mm/day).
A): zonally averaged over the continents; B): zonally averaged over the oceans

Latitude-longitude distribution of the model response

The map of surface temperature difference between the 115 kyr and 125 kyr simulations (Fig. 11) shows that even in the annual mean this response varies strongly according to geographical position. We see a region of higher temperatures over South-East Asia, over the Sahara across North America and over Northern Europe. Colder areas are found from the Mediterranean to Siberia, on the West Coast of North America and over Labrador and Canada. This latter fact is of particular interest since it could support the theories of instant glacierization that have been put forward to explain the formation of the Laurentide ice cap from extensive snow cover (19).

This decrease in annual temperature over this area is accompanied by an increase of soil water content (Fig. 12) reflecting a more positive balance of precipitation versus evaporation. This is too a favorable condition for the accumulation of an ice cover.

In sea-level pressure (Fig. 13) we have more anticyclonic conditions in January over Canada, which tend to increase the northerly flow over the eastern part and give colder temperatures (3° at 850 mb) over Labrador (Fig. 14). In July the colder temperatures (more than 5°C) in this region are mainly a response to the insolation deficit somewhat amplified over Labrador by higher cloudiness (Fig. 15). The decrease of cloudiness in January over this area leads to a surface cooling by infra-red radiation loss which acts as a positive feedback to maintain anticyclonic conditions.

CONCLUSION

It seems quite remarkable that the change of the insolation conditions alone from 125 kyr to 115 kyr values has produced in the model a cooling over the Northern high latitudes and particularly over Eastern Canada. Such result seems to support the hypothesis that warm oceans and a deficit of summer insolation can be the prime conditions for initiation of ice sheets' growth in the Northern Hemisphere (20).

We cannot however draw more definite conclusions from only a 1-year simulation since, in addition to inherent limitations of our model (low resolution, various simplifications in physical parameterizations), we cannot include the feedback of ice accumulation which operates on much longer time-scales. Furthermore the changes of sea-surface temperature which probably play an equally important role have not been taken into account in this study. Other simulations with different boundary condi-

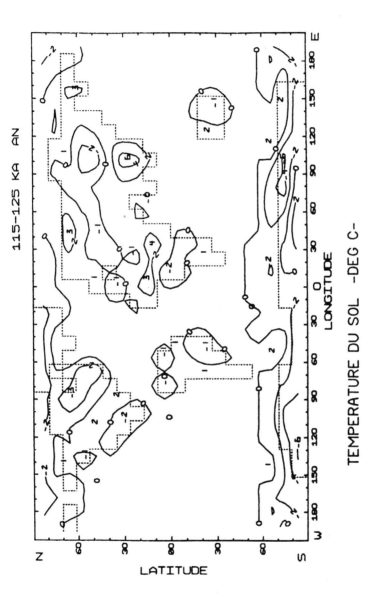

Figure 11 Geographical distribution of the difference, between the 115 kyr and 125 kyr simulations of the annual mean surface temperature (in °C). (Royer et al., 21)

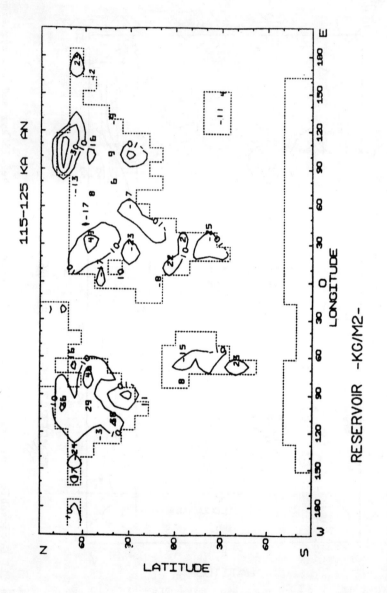

Figure 12 Geographical distribution of the difference between the 115 kyr and 125 kyr simulations of the annual mean soil water content (in kg/m^2). (Royer et al., 21)

SENSITIVITY EXPERIMENT TO ASTRONOMICAL FORCING WITH A SPECTRAL GCM 759

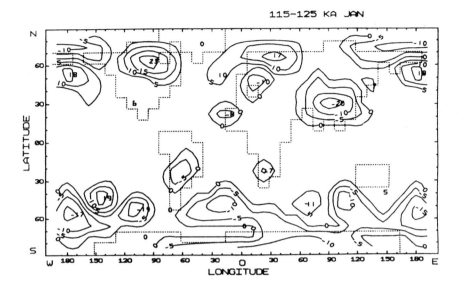

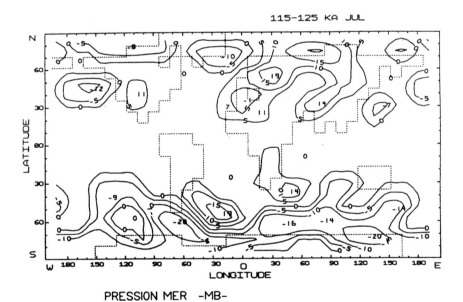

PRESSION MER -MB-

Figure 13 Geographical distribution of the difference (115 kyr minus 125 kyr) of mean sea level pressure (in mb, or hPa). A): January mean;   B): July mean

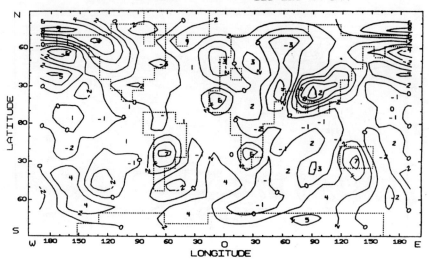

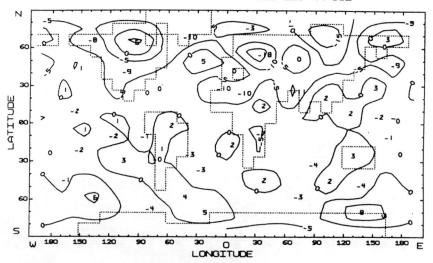

Figure 14 Geographical distribution of the difference (115 kyr minus 125 kyr) of 850 mb temperature (in °C).
A): January mean;   B): July mean

SENSITIVITY EXPERIMENT TO ASTRONOMICAL FORCING WITH A SPECTRAL GCM   761

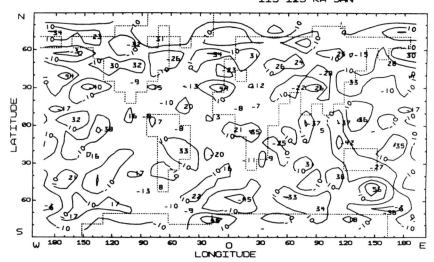

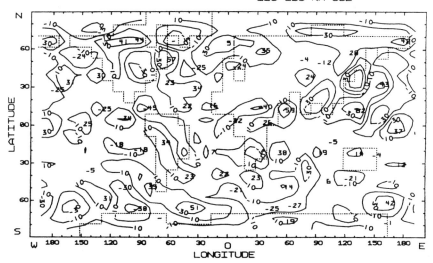

NEBULOSITE -PC-

Figure 15 Geographical distribution of the difference (115 kyr minus 125 kyr) of cloudiness (in %).
A): January mean;   B): July mean

tions and insolation conditions will be necessary in order to understand more clearly the potential influence of all these factors.

REFERENCES

1. Berger, A. : 1976, Astron. Astrophys. 51, pp. 127-135.
2. Imbrie, J. : 1982, Icarus 50, pp. 408-422.
3. Wetherald, R.T., and Manabe, S. : 1975, J. Atmos. Sci. 32, pp. 2044-2059.
4. Kutzbach, J.E., and Otto-Bliesner, B.L. : 1982, J. Atmos. Sci. 39, pp. 1177-1188.
5. Berger, A. : 1979, Il Nuovo Cimento 2Cn°1, pp. 63-76.
6. Ruddiman, W.F., and McIntyre, A. : 1977, J. Geophys. Res. 82, pp. 3877-3887.
7. Schnitker, D. : 1974, Nature 248, pp. 385-387.
8. Ruddiman, W.F., and McIntyre, A. : 1979, Science 204, pp. 173-175.
9. Royer, J.F., Deque, M., Canetti, H.J., and Boulanger, M. : 1981, "Présentation d'un modèle spectral de circulation générale à faible résolution. Simulation du climat de Janvier". Note de travail de l'EERM n°16 (internal note). Direction de la Météorologie.
10. Bourke, W. : 1974, Mon. Wea. Rev. 102, pp. 687-701.
11. Rochas, M., Rochas, G., Mittelberger, J.P., Royer, J.F., and Ernie, Y. : 1980, "Présentation d'un modèle spectral de simulation des mouvements atmosphériques à grande échelle". Note technique de l'EERM n°76. Nouvelle série. Direction de la Météorologie.
12. Royer, J.F. : 1980, "Présentation d'un modèle spectral de circulation générale à faible résolution. Expériences préliminaires avec un chauffage newtonien". Note technique de l'EERM n°83. Nouvelle série. Direction de la Météorologie.
13. Lepas, J., Le Goff, G., De Moor, G., Musson Genon, L., Pierrard, M.C., Rocafort, J.P., and Rousseau, D. : 1979, Projet Améthyste, cahier n°2, le modèle de prévision partie physique. Note technique de l'EERM n°37. Nouvelle série. Direction de la Météorologie.
14. Coiffier, J., Lepas, J., and Rivals, S. : 1981, "The New French Operational Ten-layer P.E. Model - Results and Developments", Proceedings of the Symposium on Current Problems of Weather Prediction, Vienna, June 23-26, 1981.
15. Berger, A. : 1978, J. Atmos. Sci. 35, pp. 2362-2367.
16. Alexander, R.C., and Mobley, R.L. : 1976, Mon. Wea. Rev. 104, pp. 143-148.
17. Jouzel, J., Merlivat, L., and Lorius, C. : 1982, Nature 299, pp. 688-691.

18. Chervin, R.M., and Schneider, S. : 1976, J. Atmos. Sci. 33, pp. 405-412.
19. Williams, L.D. : 1978, Quaternary Research 10, pp. 141-149.
20. Ruddiman, W.F., McIntyre, A., Niebler-Hunt, W., and Durazzi, J.T. : 1980, Quaternary Research 13, pp. 33-64.
21. Royer, J.F., Déqué, M., and Pestiaux, P. : 1983, Nature 304, pp. 44-46.

THE RESPONSE OF A CLIMATE MODEL TO ORBITAL VARIATIONS

W. D. Sellers

The University of Arizona, Institute of Atmospheric
Physics, Tucson, AR 85721, USA

ABSTRACT

A three-dimensional global climate model is used to study the response of the earth-atmosphere system to variations in the earth's orbit about the sun.

The model is basically a simplified, coarse-grid (10° latitude by 10° longitude), general circulation model with five layers in the vertical, three in the troposphere and two in the stratosphere. The dynamics are simple enough that a 5-day time step can be used. The model contains a complete hydrologic cycle, including variable snow and ice cover, soil moisture, and cloud cover, and an interactive ocean with diffusive heat transport, but no currents.

Three different sets of orbital parameters are used in computer runs of 20 model years. These parameters are those for the present (control), 9 000 years ago, and 231 000 years ago. The latter two times correspond, respectively, to warm (climatic optimum) and cold (ice age) periods on Earth.

The results indicate that land-sea temperature differences, especially in the tropics, are very sensitive to changes in the earth's orbit about the sun. The greater the summer insolation, the greater the temperature difference and the stronger and more dependable the monsoon circulation and its attendant summer rainfall. These results are in close accord with those of Kutzbach (1).

The model snow cover and sea ice area vary only slightly with changes in the orbital parameters, possibly because of defficiencies in the model.

INTRODUCTION

For a number of years the author has been trying to develop a global climate model that would have the resolution of a coarse-grid general circulation model but would not require an enormous amount of computer storage and time for an extended period run. The task is continuing. The model as it stands now and as described briefly in next section reproduces many of the major features of the global general circulation and the hydrologic cycle. However, ther are still disturbing defficiencies, which may or may not be eliminated with further research.

Several 20 model-year runs have been made using orbital parameters for the present (control run) and for 9 000 and 231 000 years ago, periods when the summer insolation at 65N was, respectively, 8.4 percent higher and 6.9 percent lower than it is now (2). The latter period was one during which, according to Budyko (3), the summer half-year temperature at 65N was 7.1K lower than the present. Kutzbach (1) studied the 9 000 BP (before the present) period and concluded that at that time the large-scale monsoon circulation of Asia was more intense and summer precipitation over land was greater (by 8 percent) than it is now.

The results of the control run are then compared briefly with observed data and also with data generated from 20-years runs with the solar constant increased and decreased by 2 percent of its present value. The effect on the model of changing the earth's orbit is discussed in the last section.

THE CLIMATE MODEL

The model uses a box grid with dimensions by 10° of latitude. There are five atmospheric layers, three in the troposphere and two in the stratosphere. A sigma (σ) vertical coordinate system is used, with the tropopause pressure (at the level $\sigma = 0$) expressed as a function of the surface temperature. The two stratospheric layers are assumed to be in radiative equilibrium, modified slightly by cooling or heating due to prescribed poleward heat transport. A five-day time step is used.

Starting with an initial three-dimensional temperature field in the troposphere, the distributions of sea-level pressure and the surface wind are determined by assuming that (a) the flow in the lowest tropospheric layer, about 1 km deep, is given by the equation of motion, retaining only the coriolis, pressure gradient, and frictional forces, (b) the flow in the upper two tropospheric layers is geostrophic and is determined by the thermal wind equation, and (c) the mean flow in the troposphere plus stratosphere is non-divergent. The stratospheric divergence is specified.

These assumptions seem to give realistic results, even though the divergence of the geostrophic wind is a very poor approximation to the divergence of the actual wind. Several attempts to use the gradient wind gave unstable results.

The vertical distribution of moisture and the precipitation rates are determined next. The surface specific humidity over land is obtained from the surface temperature and soil moisture content. A power law variation of the specific humidity with pressure is assumed in each of the three tropospheric layers. The moisture surplus, if any, in each layer is determined from the moisture continuity equation. The excess moisture is allowed either to condense or to diffuse vertically to the next higher layer.

In each layer the cloud cover is empirically related to the ratio of the condensation rate to the precipitable water vapor and to the surface relative humidity. The infrared cooling and solar heating rates of each layer are determined by considering only absorption and emissions by clouds, water vapor, carbon dioxide, and ozone. The effect of aerosol is not included. Infrared radiation is treated using a relatively simple layer emissivity scheme and solar radiation using a modified version of the Lacis-Hansen (4) model.

Over land, the soil moisture content is determined using the water balance equation, relating the time rate of change in moisture storage to the precipitation, evaporation, and run-off rates. The total energy available for evaporation is partitioned among evaporation, snow melt, and over the oceans, the melting of ice. The freezing rate of sea water is expressed as a function of the temperature and initial ice thickness.

Once the pressure, wind, moisture, and radiation fields and the surface characteristics corresponding to the initial temperature fields are determined, the temperature fields over land and water five days later are estimated from the time-dependent energy balance equations for the surface and for the bottom atmospheric layer. The lapse rate within the surface layer is expressed as a function of the surface temperature. The lapse rate in each of the upper two tropospheric layers is statistically related to the temperature at about 1 km and is modified annually to preserve a radiation balance at the top of the atmosphere.

The numerical calculations are carried out on a CYBER 175 computer housed in the Computer Center on the campus of the University of Arizona. One time step requires about 2.6 seconds of computer time (3.1 minutes per model year).

CONTROL RUN RESULTS

In Figures 1-11 observed and modeled fields of sea-level pressure (Figs. 1-3), sea-level temperature (Figs. 4-6), precipitation (Figs. 7-9), and snow and ice cover (Figs. 10,11) are compared. These will be discussed in more detail in a later paper. Briefly, the main results are the following.

a. Most of the major features of the sea-level pressure field are simulated quite well, although some of the centers are shifted away from their observed locations. The most important deficiency is probably the lack in the model of strong subtropical highs in the northern hemisphere in July. The equatorial trough is also not as pronounced as it should be. The southern hemisphere westerlies appear only because the stratospheric divergence is specified and, at 60S, is compensated by a low-level frictional convergence.

b. The main features of the sea-level temperature field are reproduced, although exaggerated in places. The northern hemisphere is too cold in winter and too warm in summer, especially over land. The southern hemisphere runs warm in both seasons. This model is only about half as sensitive as other models to changes in the solar constant. A 4 percent increase in the insolation gives a maximum temperature increase of 10K in high northern latitudes in January.

c. Model precipitation is quite variable, both in space and time. As observed, greatest amounts occur in the tropics. The Indian monsoon shows up but only weakly, possibly because there are no mountains in the model. The observed precipitation maximum off the southwest tip of South America in July is also present in the model. Increasing the solar constant by 4 percent results in an increase in global precipitation of about 20 percent (but a decrease in cloud cover of about 4 percent).

d. The sea ice amount tends to be greater than observed in the northern hemisphere and less than observed in the southern hemisphere. The seasonal cycle is reproduced fairly well, although the amplitude of the cycle is only about 55 percent of what it should be in both hemispheres. The southern hemisphere

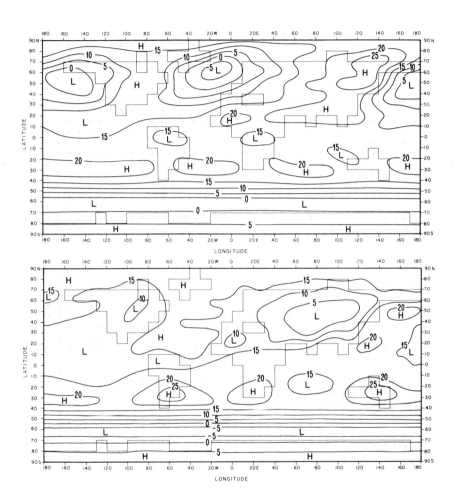

Figure 1 Sea-level pressure (mbs-1000) for January (above) and July (below) given by the climate model in year 20.

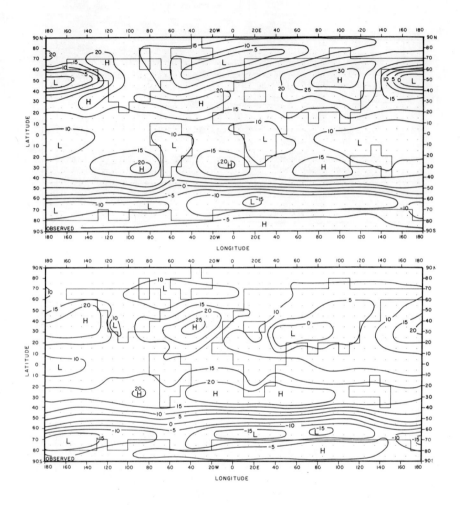

Figure 2 Observed sea-level pressure (mbs-1000) for January (above) and July (below). From Schutz and Gates (5).

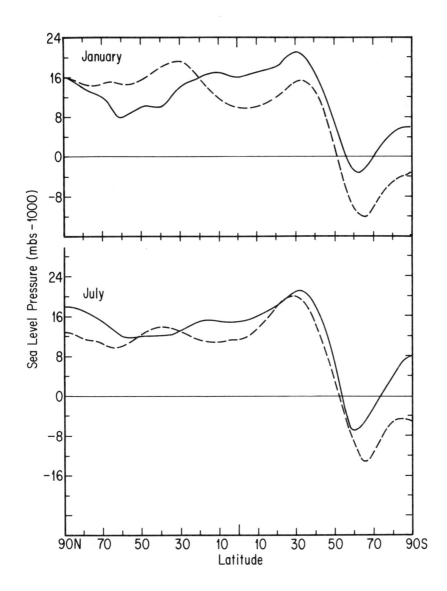

Figure 3 Zonally-averaged sea-level pressure for January and July as given by the model in year 20 (solid lines) and by observations (dashed lines ; from Schutz and Gates (5)).

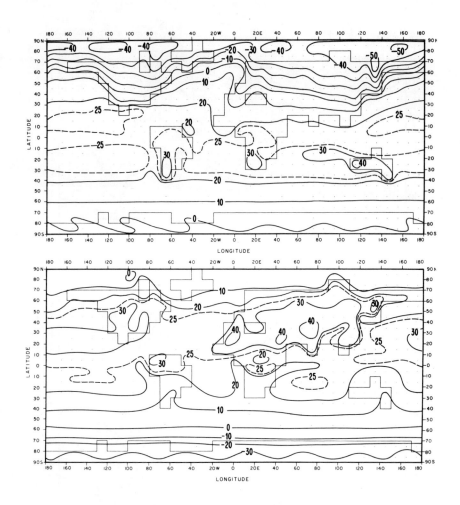

Figure 4   Sea-level temperature (C) for January (above) and July (below) given by the climate model in year 20.

THE RESPONSE OF A CLIMATE MODEL TO ORBITAL VARIATIONS 773

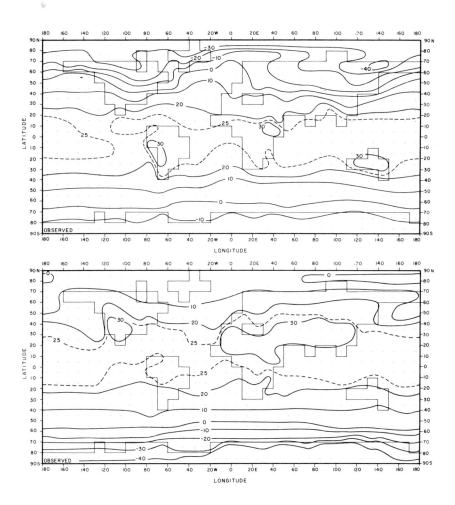

Figure 5   Observed sea-level temperature (C) for January (above) abd July (below). From Schutz and Gates (5).

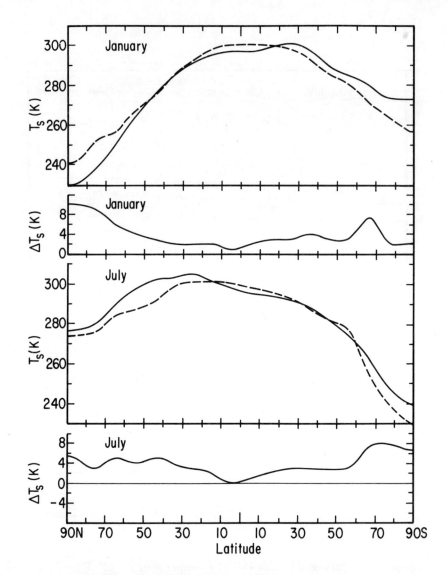

Figure 6 Zonally-averaged sea-level temperature for January and July as given by the model in year 20 (solid lines) and by observations (dashed lines ; from Schutz and Gates (5)). Also shown is the zonally-averaged temperature change obtained by the model (year 20) with an increase of the solar constant from 2 percent below to 2 percent above the present value.

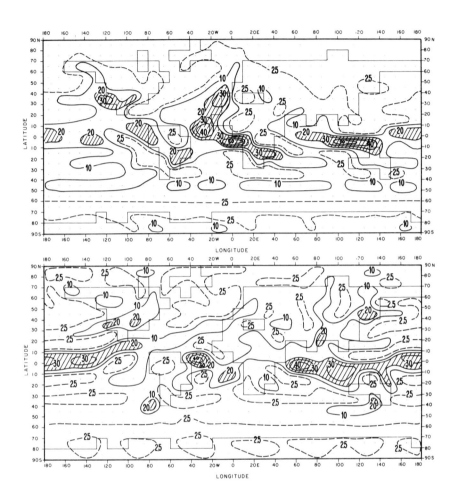

Figure 7 Precipitation (cm/month) for Dec-Feb (above) and Jun-Aug (below) given by the climate model in year 20.

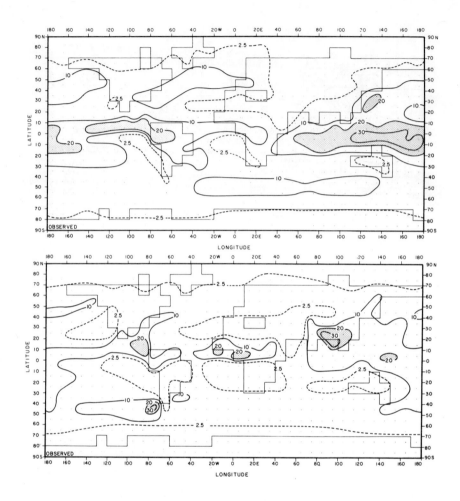

Figure 8 Observed precipitation (cm/month) for Dec-Feb (above) and Jun-Aug (below). From Schutz and Gates (5).

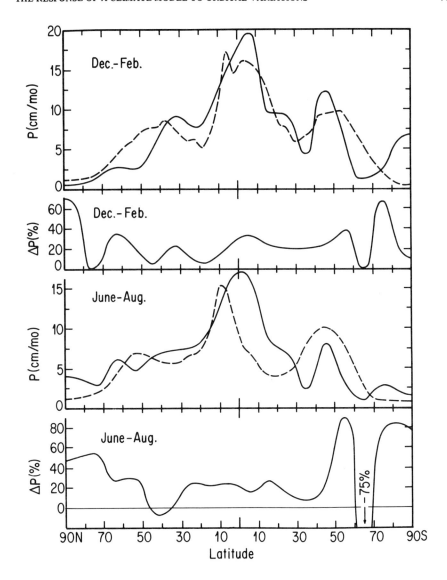

Figure 9 Zonally-averaged precipitation for Dec-Feb and Jun-Aug as given by the model in year 20 (solid lines) and by observations (dashed lines ; from Schutz and Gates (5)). Also shown is the zonally-averaged precipitation change obtained by the model (year 20) with an increase of the solar constant from 2 percent below to 2 percent above the present value.

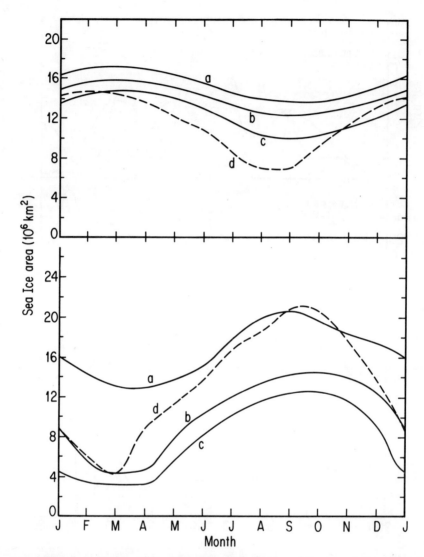

Figure 10 Annual variation of the sea-ice area in the northern hemisphere (above) and southern hemisphere (below). Curves a, b, and c are derived from the model (year 20) with the solar constant, respectively, 2 percent below, at, and 2 percent above its current value. The dashed curves (d) are based on observed data given by Newell and Chiu (6) for the northern hemisphere and by Kukla and Gavin (7) for the southern hemisphere.

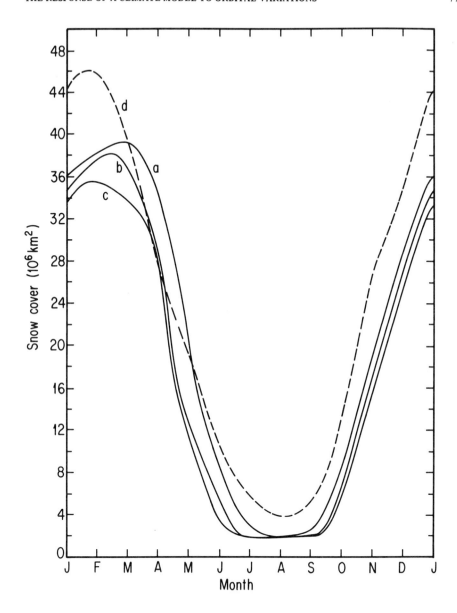

Figure 11 Annual variation of the snow cover over land in the northern hemisphere. Curves a, b, and c are derived from the model (year 20) with the solar constant, respectively, 2 percent below, at, and 2 percent above its current value. The dashed curves (d) are based on observed data given by Kukla and Kukla (8).

sea ice appears to be the most sensitive component of the model to a change in the solar constant, increasing by almost 70 percent with a two percent decrease in the insolation.

e. The extent to which the model has reached a quasi-equilibrium climate after 20 years is indicated in Figure 12, showing the time series of the differences in the annual mean global sea-level temperature and the September sea ice area in the southern hemisphere between the control run and runs with the solar constant increased and decreased by 2 percent. Although the temperature appears to be relatively steady by the end of the 20th year, the sea ice area is still changing significantly. This is especially true in the run with a low solar constant and suggests that the temperature in this case should continue to decrease slowly as time passes. Conceivably an ice-covered earth could result eventually, although this does not seem too likely.

EFFECT OF ORBITAL VARIATIONS

The effect of orbital variations on climate was studied using orbital parameters in the model for the present, 9 000 BP, and 231 000 BP. The parameters were obtained from unpublished listings provided by Andre Berger (now available as indicated in (9,10,11)).

The 9 000 BP period is of interest because, according to Kutzbach (1) it was a time of high lake levels across parts of Africa, Arabia, and India, presumably induced by enhanced summer rainfall. Similar conditions may have also existed in the southern United States (12). Kutzbach attributes the precipitation increase at least partly to an intensified continental-scale monsoon circulation, created in turn by changes in the earth's orbit about the sun.

According to Vernekar (13) summer insolation at 65N was lower 231 000 years ago than at any other time during the past or future million years. This should have been a period of widespread glaciation, as indicated, for example, by Budyko (3).

The results of this analysis are presented in Tables 1-3 and in Figures 13 and 14. In Table 1, which is similar to Table 2 of Kutzbach (1), it is seen that changing the orbital parameters had little effect on annual or global mean temperatures. However, there were significant seasonal changes, especially over the northern hemisphere continents. Compared to the present, summers were 1.9K warmer 9 000 years ago and 2.3K cooler 231 000 years ago. As indicated by the figures in parentheses in the Table, Kutzbach (1) got a warming of only 0.7K for 9 000

Table 1 Departure from the present of the sea-level temperature and precipitation for 9 000 BP and 231 000 BP. Based on model data for year 20. The figures in parentheses are from Kutzbach (1).

Departure from Present (BP)	Sea-level temperature (K)		Precipitation (cm/month)	
	9 000	231 000	9 000	231 000
June to August				
Northern Hemisphere, land	1.9(0.7)	-2.3	2.36(1.2)	-0.53
Southern Hemisphere, land	1.2(0.3)	-0.6	0.85(0.6)	0.20
Global, land and ocean	0.6(0.2)	-0.8	0.53(0.0)	-0.41
Annual				
Northern Hemisphere, land	-0.7(-0.2)	-0.1	0.26(0.6)	-0.12
Southern Hemisphere, land	0.1(0.2)	-0.7	-0.55(-0.6)	-0.23
Global, land and ocean	-0.2(0.0)	-0.1	0.08(0.0)	-0.08

Table 2 Northern Hemisphere snow cover (land) and sea ice and Southern Hemisphere sea ice (10^6 km^2) for March and September of year 20. Based on model data using orbital parameters for the present, 9 000 BP and 231 000 BP.

	NH snow(land)		NH sea ice		SH sea ice	
	Mar	Sep	Mar	Sep	Mar	Sep
Present	36.896	2.070	15.669	12.424	4.294	14.150
9 000 BP	40.138	2.095	15.735	10.611	3.908	13.723
231 000 BP	34.621	2.419	15.880	13.404	5.669	14.915

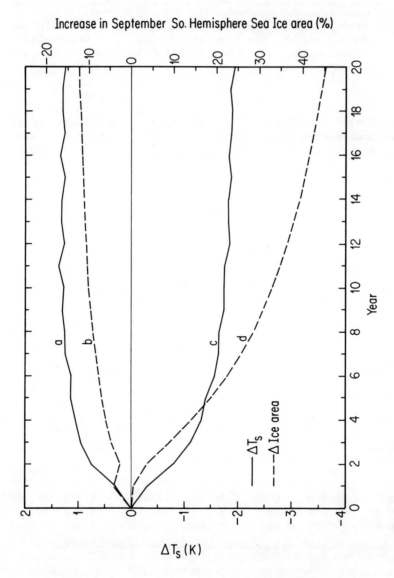

Figure 12 Variation with model time of the difference in mean annual global sea-level temperature and in southern hemisphere September sea ice area between the runs with the solar constant 2 percent above (curves a, b) and 2 percent below (curves c, d) and the control run.

BP. This may be because he assumed no change in ocean surface temperatures. However, in the model used here the northern hemisphere oceans were only 0.1K warmer in summer 9 000 years ago than they are today, so his assumption may not be a bad one.

The small changes in annually averaged temperatures are reflected in small changes in snow and ice cover (Table 2). The area covered by sea ice in both hemispheres in September was only 5 to 7 percent greater 231 000 years ago than it is now. Decreasing the solar constant by 2 percent increased sea ice by 45 percent in September in the southern hemisphere (Fig. 11) and by 10 percent in the northern hemisphere. In this model sea ice is not very sensitive to changes in the orbital parameters. Because of the cooler winters and in spite of slightly reduced precipitation the model gives more snow cover, by about 9 percent, over the northern hemisphere land masses in winter 9 000 years ago than there is today.

Returning to Table 1, the results of this study confirm those of Kutzbach (1), who obtained a significant increase in summer monsoon rainfall over the northern hemisphere continents 9 000 years ago. The magnitude of the increase is about doubled by the model used here. A less significant decrease occurred 231 000 years ago.

In Figure 13 is shown the latitudinal variation of June to August precipitation zonally-averaged over land for years 11-20 for each of the orbital runs. Error bars are included, when the standard error exeeds 0.1 cm/month, to give some indication of the significance of the differences shown. The greatest increases in summer precipitation 9 000 years ago, approaching 50 percent, obviously occurred in the tropics, with a secondary maximum between 30 and 40N. The stronger monsoon circulation at that time was driven by a 50 percent increase, from 3.6K to 5.4K, in the summer temperature difference between land and water averaged over the entire northern hemisphere. Similarly 231 000 years ago the temperature difference was 50 percent less and the monsoon circulation correspondingly weaker. Year-to-year variability of precipitation in the tropics was moderately high, but not enough so to detract from the significance of the results.

In Figure 14 is shown the difference in June to August precipitation between 9 000 BP and the present, based on data for 20 of the model runs. Since the figure is for only one summer and since there is considerable temporal variability, especially in the tropics, the centers of maximum precipitation change should be expected to vary slightly in location and magnitude from year to year, but not enough to affect the overall results.

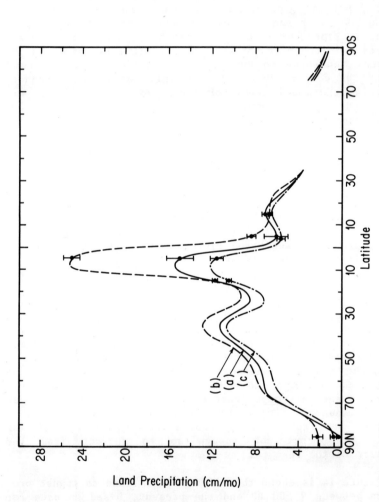

Figure 13 The latitudinal variation of Jun-Aug precipitation zonally-averaged over land for years 11-20 for runs with the orbital parameters for (a) the present, (b) 9 000 years ago, and (c) 231 000 years ago. Error bars are given for latitudes at which the standard error of the mean exceeds 0.1 cm/month.

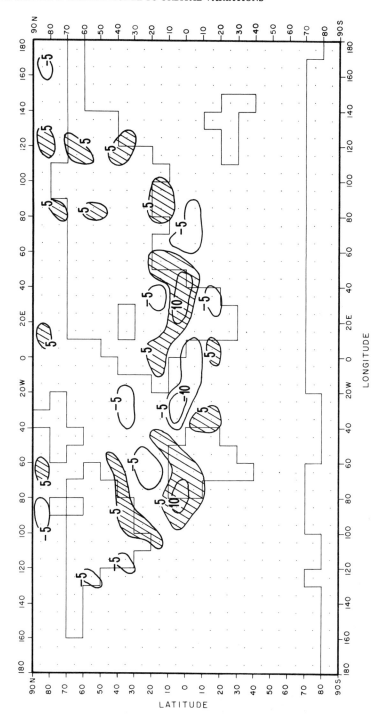

Figure 14 Difference in Jun–Aug precipitation (cm/month) in year 20 between the runs with orbital parameters for 9 000 years ago and for the present.

It is encouraging that some of the centers of maximums precipitation increase in Figure 14 occur where paleoclimatic evidence suggests heavier precipitation 9 000 years ago than today; i.e., Africa, Arabia, India, and the southern United States. The model also indicates heavier summer precipitation than today over northern South America and parts of eastern Asia and less precipitation than today in the tropical and subtropical Atlantic and India oceans and along the west coast of North America.

In Table 3 is given the seasonal variation of precipitation over land and water between 0 and 10N averaged for years 11-20 of the model runs. Also shown are the standard errors of the means. Generally the heaviest precipitation in this region occurs in the spring and fall over land and in the fall over water. The major exception is a shift to a strong summer maximum over land 9 000 years ago. Relative to the present, the land masses were wetter in the spring and summer and drier in the fall and winter 9 000 years ago and vice versa 231 000 years ago. Over water the only significant change was a drier spring 9 000 years ago than today.

Table 3 Seasonal precipitation between 0 and 10N averaged for years 11-20. S.E. is the standard error of the mean. The units are cm/month

	Dec-Feb Ave	S.E.	Mar-May Ave	S.E.	Jun-Aug Ave	S.E.	Sep-Nov Ave	S.E.
LAND								
Present	11.8	1.0	20.5	1.0	15.0	1.2	19.7	1.4
9 kyr BP	9.8	0.4	21.0	0.5	25.1	0.8	16.2	1.3
231 kyr BP	15.3	0.7	15.6	0.5	11.6	0.6	21.0	0.7
WATER								
Present	15.6	0.7	15.5	0.5	16.0	0.4	17.2	0.4
9 kyr BP	15.3	0.2	13.7	0.2	16.0	0.2	17.3	0.3
231 kyr	16.1	0.3	16.3	0.4	15.2	0.2	16.3	0.3

Table 3 indicates that the year-to-year variability of tropical rainfall in all seasons is greater over land than over water, which might be expected, and that for both land and water the variability is greater today than it was either 9 000 or 231 000 years ago. This could be a quirk of the model. However, it could also be explained as a reflection of the intensity of the monsoon circulation in both summer and winter. 9 000 years ago the circulation was stronger than today and more dependable in producing summer rains and winter dryness. On the other hand, 231 000 years ago, when summer was apparently the driest season of the year, the monsoon could be counted on not to appear. This is an interesting result and one which should be checked with other models.

CONCLUSION

In summary, the climatic picture painted by this model for the northern hemisphere land masses 9 000 years ago is one of cooler, drier winters with more snowfall; warmer, wetter summers; and a stronger, more dependable monsoon circulation than today. Essentially the reverse situation existed 231 000 years ago. The model gives little evidence that changes in the orbital parameters could produce an ice age. However, there are enough uncertainties left to leave that question open. For example, as mentioned earlier and shown in Figure 6, the model temperatures in the northern hemisphere are too high in summer and too low in winter. This would tend to discourage the onset of an ice age. At this point the main effect of orbital changes seems to be to modulate seasonal land-sea temperature differences and, hence, to regulate monsoon circulations.

REFERENCES

1. Kutzbach, J. E.: 1981, Science 214, pp. 59-61.
2. Berger, A.L. : 1979, Nuovo Cimento 2C (1), pp. 63-87.
3. Budyko, M. I.: 1977, "Climatic Changes". Washington, D.C., American Geophysical Union, 261 pp.
4. Lacis, A. A., and Hansen, J. E.: 1974, J. Atmos. Sci. 31, pp. 118-133.
5. Schutz, C., and Gates, W. L.: 1971, 1972,"Global climatic data for surface, 800 mb, 400 mb: January, July." R-915, 1029-ARPA, Santa-Monica, The Rand Corporation, 173 pp. 180 pp.
6. Newell, R. E., and Chiu, L. S.: 1981, in: "Climatic Variations and Variability: Facts and Theories", A. Berger (Ed.), Boston, D. Reidel Publ. Co., pp. 21-61.
7. Kukla, G. J., and Gavin, J.: 1981, Science 214, pp. 497-503.
8. Kukla, G. J., and Kukla, H. J.: 1974, Science 183, pp. 709-714.
9. Berger, A.L. : 1982, "Numerical Values of the Elements of the Earth's Orbit from 5 000 kyr BP to 1 000 kyr AP", Contribution 35, Institut d'Astronomie et de Géophysique G. Lemaître, UCL, Louvain-la-Neuve.
10. Berger, A.L. : 1982, "Numerical Values of Mid-Month Insolations from 1 000 kyr BP to 100 kyr AP", Contribution 36,

Institut d'Astronomie et de Géophysique G. Lemaître, UCL, Louvain-la-Neuve.
11. Berger, A.L. : 1982, "Numerical Values of Caloric Insolations from 1 000 kyr BP to 100 kyr AP", Contribution 37, Institut d'Astronomie et de Géophysique G. Lemaître, UCL, Louvain-la-Neuve.
12. Cole, K.: 1982, Science 217, pp. 1142-1145.
13. Vernekar, A. D.: 1968, "Research on the Theory of Climate Volume II: Long-Period Global Variations of Incoming Solar Radiation". Hartford, Conn., The Travelers Research Center. Inc., 289 pp.

INFLUENCE OF THE CLIMAP ICE SHEET ON THE CLIMATE OF A GENERAL CIRCULATION MODEL : IMPLICATIONS FOR THE MILANKOVITCH THEORY

S. Manabe, A.J. Broccoli

Geophysical Fluid Dynamics Laboratory/NOAA, Princeton University, P.O. Box 308, Princeton, New Jersey 08540, USA

INTRODUCTION

While the spectral analysis of data from deep sea cores supports the suggestion that the temporal variation of the earth's orbital parameters is responsible for triggering the growth and decay of continental ice during the Quaternary (1), the physical factors responsible for maintaining a global ice age climate are less well understood. One of the important factors in maintaining a lowered atmospheric temperature during an ice age is the presence of large continental ice sheets that reflect a large fraction of incoming solar radiation. This study investigates the influence of the 18 kyr BP (18 000 years before present) ice sheet on the earth's climate by use of a mathematical model of climate constructed at the Geophysical Fluid Dynamics Laboratory of NOAA.

As the box diagram of Figure 1 indicates, the mathematical model used for this study consists of three basic units : 1) a general circulation model of the atmosphere, 2) a heat and water balance model over the continents, and 3) a simple model of the oceanic mixed layer. A brief description of these three units follows. A more detailed description of this atmosphere-mixed layer ocean model can be found in Manabe and Stouffer (2).

The atmospheric general circulation model computes the rates of change with time of the vertical component of vorticity, horizontal divergence, temperature, moisture, and surface pressure using the so-called "spectral method". The dynamical component of this model is developed by Gordon and Stern (3). Manabe et al. (4) and Manabe and Hahn (5) discuss

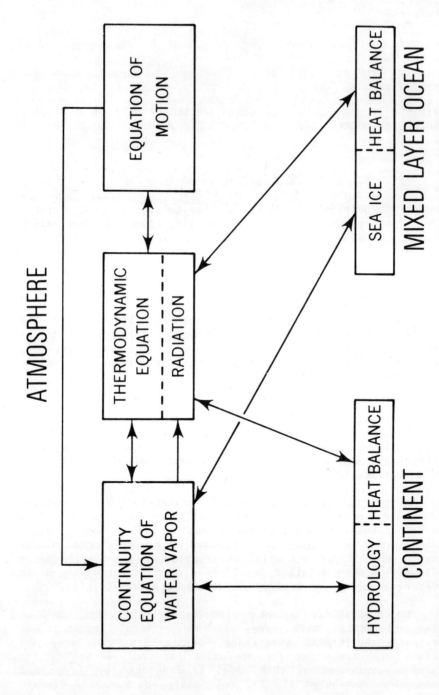

Figure 1 Box diagram illustrating the basic structure of the mathematical model of climate.

the structure and performance of this atmospheric model in detail.

Over the continents, the assumption of zero surface heat storage is used to determine surface temperatures from energy fluxes at the surface. Snow is allowed to accumulate on the surface, with the change in snow depth predicted as the net contribution from snowfall, sublimation, and snowmelt. A higher surface albedo is used when snow is present. Also used is a water balance model which computes changes in soil moisture from the rates of rainfall, evaporation, snowmelt and runoff. Further details of the hydrologic computations can be found in Manabe (6).

The oceanic mixed layer model consists of a vertically isothermal layer of static water of uniform depth. This model includes the effects of oceanic evaporation and the large heat capacity of the oceanic mixed layer, but neglects the effects of horizontal heat transport by ocean currents and of the heat exchange between the mixed layer and the deeper parts of the ocean. Sea ice is predicted when the mixed layer temperature falls below the freezing point of sea water (-2°C), and a higher surface albedo is used where sea ice is present.

EXPERIMENTAL DESIGN

In order to investigate the influence of continental ice sheets on the climate of an ice age, two long term integrations of the atmosphere-mixed layer ocean model described in the preceding section are conducted. The first time-integration, hereafter identified as the standard experiment, assumes as a boundary condition the modern distribution of continental ice. The second time-integration assumes the distribution of continental ice at the time of the last glacial maximum as reconstructed by CLIMAP (7), and will hereafter be identified as the ice sheet experiment.

A sea level difference between the two experiments of 150 m is prescribed, consistent with the glacial lowering of sea level as estimated by CLIMAP. Orbital parameters in both experiments are set at modern values, making the distribution of insolation at the top of the atmosphere with latitude and season the same in both experiments. This simplification is reasonable, since the orbital parameters at 18 kyr BP are not very different from the modern values. The surface albedo distribution of snow- and ice-free areas is prescribed to be the same in both experiments.

The initial condition for both time integrations is a dry, isothermal atmosphere at rest coupled with an isothermal mixed layer ocean. In both cases, the model is time integrated for 20 seasonal cycles. A quasi-equilibrium model climate is achieved after 15 model years, and the subsequent five-year period is used for analysis in each experiment.

In Figure 2, the geographical distribution of February surface air temperature of the model atmosphere is compared with the surface distribution compiled by Crutcher and Meserve (8) and Taljaad et al. (9). This comparison indicates that, despite some exceptions, the model succeeds in reproducing the general characteristics of the observed distribution of surface air temperature. The success of the model in simulating the geographical distribution of climate and its seasonal variation encouraged the authors to conduct the present study by use of this model.

THERMAL RESPONSE

Sea Surface Temperature

To illustrate the effect of continental ice on the distribution of sea surface temperature (SST), Figures 3 and 4 are constructed. These figures show the geographical distributions of the SST difference of the model mixed layer ocean between the ice age and standard experiments for February and August, respectively. These can be compared to the distributions of SST difference between 18 kyr BP and the present as determined by CLIMAP, which are added to the lower half of each figure.

In the Northern Hemisphere in both winter and summer, SSTs from the ice sheet experiment are significantly lower than the corresponding temperatures in the standard experiment. The SST differences are most pronounced in the mid-latitudes of the North Atlantic and the North Pacific, with the ice sheet-induced cooling over the Atlantic generally larger than the corresponding cooling over the Pacific. This is in good qualitative agreement with the difference between the modern and 18 kyr BP distributions of SST as obtained by CLIMAP.

In contrast, the SST differences in the Southern Hemisphere of the model are very small in both seasons, while the SST differences as estimated by CLIMAP are of significant magnitude. Since most of the changes in the distribution of continental ice between the two experiments are located in the Northern Hemisphere, this result suggests that the presence of an ice sheet in one hemisphere has relatively little influence on the distribution of sea surface temperature in the other hemisphere.

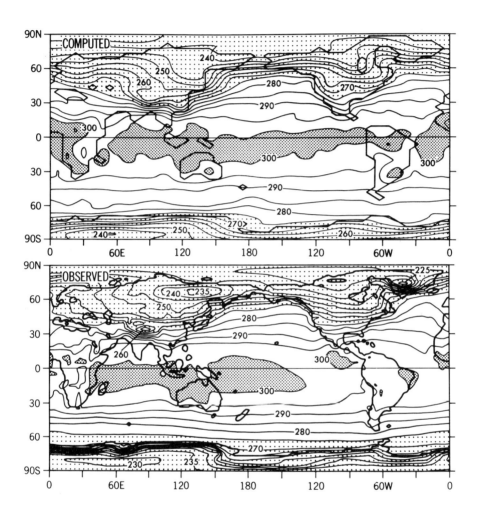

Figure 2 Geographical distributions of monthly mean surface air temperature (degrees Kelvin) in February. Top : computed distribution from the standard experiment. Bottom : observed disitribution (8,9). The computed surface air temperature represents the temperature of the model atmosphere at the lowest finite difference level located at about 70 m above the earth's surface.

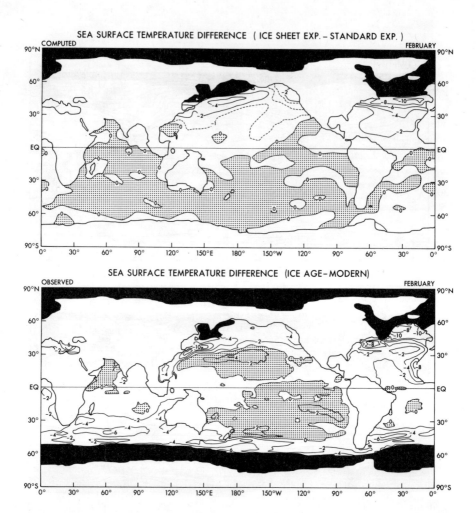

Figure 3 Geographical distribution of February sea surface temperature differences (degrees Kelvin). Top : difference between ice sheet and standard experiments. Bottom : difference between 18 kyr BP and present (as reconstructed by CLIMAP).

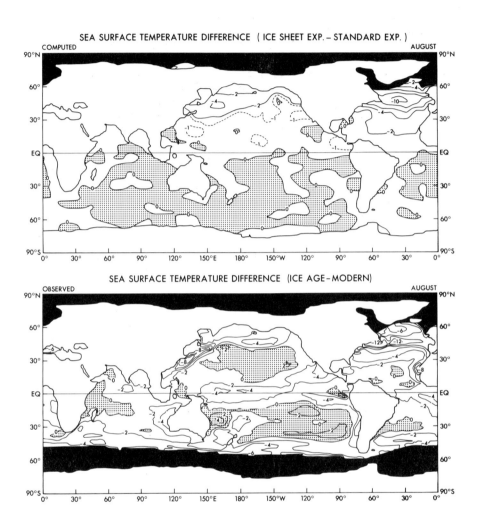

Figure 4 Same as Figure 3 except for August.

Hemispheric Heat Balance

To evaluate the lack of a Southern Hemisphere response to the presence of widespread continental ice in the Northern Hemisphere, the hemispheric heat budgets of the model atmosphere are obtained from the standard and ice sheet experiments. These are illustrated by the box diagrams in Figure 5, as is the difference in heat budget between the two experiments.

For the Northern Hemisphere, the net downward flux of solar radiation at the top of the model atmosphere in the ice sheet experiment is 5.6 W m^{-2} less than the corresponding flux in the standard experiment. This is primarily due to the reflection of insolation by the large area of continental ice. The difference in incoming solar radiation is essentially counterbalanced by a difference of 5.8 W m^{-2} in outgoing terrestrial radiation at the top of the model atmosphere. Although the rate of interhemispheric heat exchange is also different between the two experiments, the magnitude of the difference is only 0.4 W m^{-2}, and is much smaller than the differences in the net incoming solar radiation and net outgoing terrestrial radiation.

These results indicate that, in the ice sheet experiment, the effective reflection of incoming solar radiation reduces the surface and atmospheric temperatures in the Northern Hemisphere of the model and, accordingly, the outgoing terrestrial radiation at the top of the atmosphere (10). The relatively low surface temperature in the Northern Hemisphere induces a small increase in the heat supplied from the warmer atmosphere in the Southern Hemisphere. However, the radiative compensation in the Northern Hemisphere is much more effective than the thermal adjustment through the interhemispheric heat exchange in the model atmosphere.

CONCLUSIONS

The results of this study suggest that the effects of increased continental ice extent alone are insufficient to explain the glacial climate of the Southern Hemisphere, although variations in the earth's orbital parameters may be responsible for including the large fluctuations in the extent of Northern Hemisphere ice sheets during the Quaternary. Thus it is necessary to look for mechanisms other than the interhemispheric exchange of heat in the atmosphere in order to explain the low temperature of the Southern Hemisphere during the 18 kyr BP ice age. This is consistent with the results of Suarez and Held (11) using a simple energy balance model to study the astronomical theory of the ice ages.

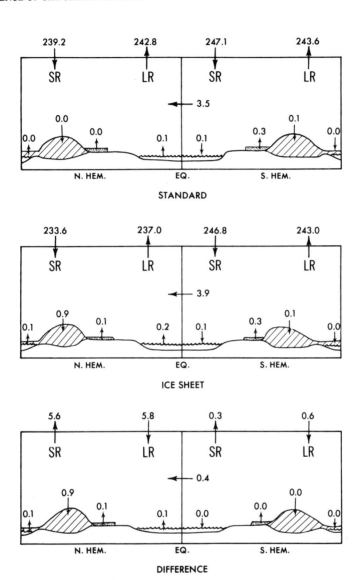

Figure 5 Box diagrams representing the primary energy fluxes in the atmosphere-ocean-cryosphere system of the model. The surface fluxes shown at the bottom of the Northern Hemisphere box represent the heat energy involved in the growth and decay of sea ice, land ice, and permanent snowcover, and long-term changes in heat storage in the oceanic mixed layer. Top : standard experiment. Center : ice sheet experiment. Bottom : difference between ice sheet and standard experiments.

Among the potential mechanisms for the cooling of the Southern Hemisphere during glacial times is the cross-equatorial heat transport by the ocean circulation. For example, if the intensity of the interhemispheric thermohaline circulation changes from a glacial to an interglacial period, as is suggested by Rooth (12), it is possible that the interhemispheric ocean heat transport does also, resulting in a change of Southern Hemisphere temperature.

Other processes which can cause an almost simultaneous change of temperature in both hemispheres are fluctuations in the concentration of carbon dioxide of the loading of aerosols in the atmosphere. Indeed, the results from the recent analysis of ice cores from the Antarctic and Greenland ice sheets suggest that the atmospheric concentration of CO_2 during the last glacial maximum was about 200 ppm by volume and is significantly less than the current concentration of 340 ppm (13,14). Broecker (15) proposes a mechanism by which a reduction of atmospheric carbon dioxide might result from the lowering of sea level associated with the growth of continental ice. If such reduction did occur, it could account for much of the Southern Hemisphere cooling indicated in the CLIMAP results.

REFERENCES

1. Hays, J.D., Imbrie, J., and Shackleton, N.J. : 1976, Science 194, pp. 1121-1132.
2. Manabe, S., and Stouffer, R.J. : 1980, J. Geophys. Res. 85, pp. 5529-5554.
3. Gordon, C.T., and Stern, W.F. : 1982, Mon. Wea. Rev. 110, pp. 625-644.
4. Manabe, S., Hahn, D.G., and Holloway, J.L. : 1979, GARP Publ. Ser. 22, World Meteorological Organization, Geneva.
5. Manabe, S., and Hahn, D.G. : 1981, Mon. Wea. Rev. 109, pp. 2260-2286.
6. Manabe, S. : 1969, Mon. Wea. Rev. 97, pp. 739-774.
7. CLIMAP Project Members : 1981, The Geological Society of America Map and Chart Series, MC-36.
8. Crutcher, H.L., and Meserve, J.M. : 1970, NAVAIR 50-IC-52, U.S. Naval Weather Service, Washington, D.C.
9. Taljaad, J.J., van Loon, H., Crutcher, H.L., and Jenne, R.L. : 1969, NAVAIR 50-IC-55, U.S. Naval Weather Service, Washington, D.C.
10. Bowman, K. : 1982, J. Geophys. Res. 87, pp. 9667-9674.
11. Suarez, M.J., and Held, I.M. : 1979, J. Geophys. Res. 87, pp. 9667-9674.
12. Rooth, C. : 1982, Prog. Oceanog. 11, pp. 131-149.

13. Berner, W., Oeschger, H., and Stouffer, B. : 1980, Radiocarbon 22, pp. 227-235.
14. Delmas, R.J., Ascencio, J.M., and Legrand, M. : 1980, Nature 284, pp. 155-157.
15. Broecker, W. : 1982, Prog. Oceanog. 11, pp. 151-197.

[1] Strickly speaking, the ice sheet-induced reduction of surface temperature results not only from the large surface reflectivity but also from the high elevation of the ice sheet surface.

THE SENSITIVITY OF MONSOON CLIMATES TO ORBITAL PARAMETER CHANGES FOR 9 000 YEARS BP : EXPERIMENTS WITH THE NCAR GENERAL CIRCULATION MODEL

J.E. Kutzbach, P.J. Guetter

IES-Center for Climatic Research, University of Wisconsin-Madison, Madison, Wisconsin 53706, USA

ABSTRACT

The July and January climates of 9 000 years ago are simulated with a high resolution general circulation model. The incoming solar radiation is specified from the orbital parameters for 9 000 years ago, and the effect of the remnant North American ice sheet is taken into account (sea-surface temperatures are not changed from modern values). The change in radiation causes temperatures to be higher in July and lower in January over large areas of the continents ; the increased temperature contrast between land and ocean results in intensified monsoon circulation and increased precipitation over tropical lands. These findings agree with paleoclimatic evidence and with the results of experiments done using a low resolution model.

INTRODUCTION

Milankovitch (1) showed that variations of the Earth's orbital parameters (obliquity, eccentricity, and time of perihelion) could have caused major glacial/interglacial changes of climate. The geologic record of the past million years obtained from ocean sediment cores, contains evidence of major climatic changes that agree in their timing with the orbital parameter changes (2). Moreover, studies with time-dependent climate models have confirmed and clarified links between orbital changes and ice ages (3,4).

In previous studies by Kutzbach (5) and Kutzbach and Otto-Bliesner (6), sensitivity experiments with a low-resolution general circulation model were used to show that changes of the seasonal cycle of solar radiation, produced by changes of the Earth's orbital parameters, could have been an important factor in causing the dramatic climatic changes of the early Holocene, in particular, the intensified monsoon circulations that have been inferred from the geologic record (7,8). This paper continues to explore the response of monsoonal climates to orbital parameter changes.

The seasonal cycle of solar radiation at 9 kyr BP (Before Present) differed strikingly from that of the present. The tilt of the earth's rotational axis was larger (24.24° at 9 kyr BP; 23.44° at present) and perihelion occurred in Northern Hemisphere summer rather than in winter (30 July at 9 kyr BP; 3 January at present); there was also a small change in the eccentricity of the earth's orbit. These three changes, when incorporated into the solar radiation equations, produced solar radiation differences (9 kyr BP minus present) that are large at all latitudes of the Northern Hemisphere; about 7% more radiation in June-July-August and about 7% less radiation in December-January-February, as was also shown in (19).

The response of the surface temperature of the ocean to the altered seasonal radiation cycle should have differed considerably from that of the land (5). The large heat storage capacity (thermal inertia) of the ocean's mixed layer should have damped the seasonal response of the ocean temperature to the amplified solar radiation cycle. A rough estimate was that the ocean temperature change for a 100 m mixed layer should have been of order ± 0.5 K (i.e., warmer in July, colder in January) whereas the land surface temperature change in continental interiors of the Northern Hemisphere could have been an order of magnitude larger (± 5 K; warmer in July, colder in January). This critical difference in thermal response of ocean and land appears to be supported by the limited amount of paleoclimatic evidence. With the exception of certain coastal areas, the ocean surface temperature at 9 kyr BP appears to have been very similar to present conditions (Ruddiman, Prell, 1983, personal communication). In contrast, the record of climate on land suggests increased seasonality or continentality at 9 kyr BP (6,9).

When the seasonal cycle of solar radiation for 9 kyr BP was used to drive the low-resolution general circulation model, and when the ocean surface temperature was prescribed to have its present seasonal cycle, an intensified monsoon circulation was simulated. The surface temperature for June to August 9 kyr BP was 2-4 K higher than for the modern simulation over much of

Eurasia, the surface pressure difference between land and ocean increased (lowered pressure over the warmer land), and both the Asian monsoon currents and the cross-equatorial flow from the Southern to the Northern Hemisphere over Africa and the western Indian Ocean intensified considerably (see Figures 1-5;(6)). Precipitation and precipitation-minus-evaporation were increased over North Africa and the Middle East and South Asia. During December to February, the Eurasian land-mass was colder than for the modern simulation and the winter monsoon was slightly stronger.

The primary purpose of this paper is to report the results of similar experiments made with the Community Climate Model (CCM) of the National Center for Atmospheric Research (NCAR). This model has higher horizontal and vertical resolution and more realistic parameterization of physical processes than the low-resolution model (LRM) used in the previously reported 9 kyr BP experiment. Because of the higher spatial resolution of the CCM compared to the LRM, the simulated pattern of climate contains greater detail that will facilitate comparison with the geologic record. Moreover, the improved physical parameterizations of cloudiness-radiation processes in the CCM will permit changes in cloudiness, radiation, and energy budgets to be examined.

The primary conclusion gained from this inter-model comparison will be that the general patterns of the 9 kyr BP climate are simulated identically by both the CCM and the LRM. The CCM simulates somewhat greater changes. These new results based upon experiments with the CCM therefore further support the findings that were reported previously concerning the important role of orbital parameter changes, associated with the differential thermal response of continents and oceans, in either enhancing or weakening the large-scale monsoon (seasonal) circulations of the past.

THE NCAR CCM

A description of the NCAR CCM and its simulation of the modern January and July climate is provided by Pitcher et al. (10). Both the CCM and LRM incorporate atmospheric dynamics based upon the equations of fluid motion and include radiative and convective processes, condensation and evaporation. A surface heat budget is computed over land. Orographic influences of mountains are also included. Ocean surface temperatures, sea ice limit, and snow cover are prescribed for either July or January conditions. Some of the differences between the CCM and the LRM are summarized here; for details concerning each model, refer to Pitcher et al. (10) and Otto-Bliesner et al.

(11). The CCM has more vertical levels than the LRM (9 versus 5) and a higher horizontal wavenumber and therefore higher effective latitude-longitude resolution (4.4° by 7.5° versus 11.6° by 11.2° latitude-longitude grid size). The equations used to calculate the downward and upward streams of short-wave (solar) and long-wave (terrestrial) radiation are more detailed and contain more degrees of freedom in the CCM than in the LRM. The CCM calculates cloudiness as a function of the model-generated vertical motion and moisture fields; the clouds, in turn, influence the streams of short-wave and long-wave radiation. In contrast, the LRM prescribes a seasonally-varying, latitudinal average cloudiness that is independent of the calculated motion and moisture fields; therefore, the possibility of radiation cloudiness interaction is excluded from the LRM. The improved treatment of radiation in the CCM has helped to produce a fairly realistic simulation of the observed general circulation (12).

OUTLINE OF EXPERIMENTS

Perpetual July and January

The version of the CCM used here is configured for so-called perpetual July and January experiments (10); that is, the model is integrated for an extended period of, say, 90 or more simulated days, but with solar radiation held constant for January 16 or July 16 values. The solar radiation for these calender dates can be expressed equivalently in terms of degrees of celestial longitude measured from the vernal equinox. For the 9 kyr BP perpetual January and July experiments, the radiation calculations use the same celestial longitude as in the control case in order to simulate conditions for the same point in the seasonal cycle (see (13) or (6) for a discussion of "calendar date" versus "celestial longitude" perspectives).

Because the LRM used seasonally-varying solar radiation and simulated an entire annual cycle, there is no exact correspondence between the previous results and these new results for perpetual July and January. However, because the month-to-month changes of solar radiation are small near summer and winter solstices, the CCM results for perpetual July and January of 9 kyr BP will be compared with previously-published LRM results for June-July-August and December-January-February of 9 kyr BP (6). Plans for simulating an entire seasonal cycle for 9 kyr BP with the CCM have been made.

Length of experiments, comparison with control

Each experiment consisted of a 120-day simulation. The experiments were started with all model variables set at values for day 400 of an NCAR CCM control simulation but with the solar radiation changed from modern to 9 kyr BP conditions. The first 30 days of each 120-day simulation were treated as an adjustment period and ignored. The final 90 days of the simulation were then averaged and compared to a 90-day average of the control (modern) simulation.

Because of the inherent variability of the simulated climate even with no change in external conditions, a measure of the model variability (noise) is required in order to assess the statistical significance of simulated climatic changes (14). The model noise statistics were computed from a 1400-day control simulation that was made available by Blackmon (personal communication, 1982). From the final 1050 days of this simulation, eight seven 90-day segments were taken, each separated by 60 days. The seven 90-day segments were then averaged to obtain an ensemble mean. The model noise level for each variable and grid point was then estimated as the standard deviation of the seven 90-day averages from the ensemble mean. For the results reported here, a climatic difference (9 kyr BP minus Present) was statistically significant at the 5% (1%) level if the difference exceeded 3.5 (5.2) standard deviations, (two-sided t-test, 6 degrees of freedom; see (14)). The significance levels for July experiments were based upon six 90-day segments (rather than seven as was the case for January) ; and, corresponding to 5% (1%) levels, the differences had to exceed 3.6 (5.7) standard deviations.

Ice Sheet Sensitivity

It was shown previously (6) that the climatic conditions for 9 kyr BP were influenced not only by the changed seasonal cycle of solar radiation but also by the presence of the North American ice sheet. The 9 kyr BP ice sheet was considerably reduced in size and especially volume as compared to its counterpart at the glacial maximum (15,16); nevertheless, it still covered 30 grid points in the model, an area of 5.4×10^6 km^2.

In order to isolate the climate sensitivity to both the solar radiation and the ice sheet, one July experiment changed only the solar radiation to its 9 kyr BP values; a second July experiment changed solar radiation and also inserted the 9 kyr BP ice sheet. The ice sheet sensitivity should be largest in July when the high-albedo ice sheet replaces the low-albedo land surface of the control experiment over North America; in January, the area covered by the ice sheet was snow-covered in the control experiment and, in any case, the solar radiation is much less in January than in July. Therefore the ice sheet sensitivity is primarily an orographic response in January whereas the sensitivity is a response to both orographic and thermal forcing in July.

As in the LRM experiments, a maximum thickness of 800m and an albedo of 0.5 were assigned to the 9 kyr BP ice sheet. The relatively low albedo (in comparison to an albedo of 0.8 for fresh snow or ice) was used because the ice was old and may have had considerable melt water or even wind-blown soil material and vegetation on parts of its surface.

RESULTS

July (no ice sheet)

The increased solar radiation of July 9 kyr BP causes higher surface temperatures over the Northern Hemisphere continents (Fig. 1a). The increase exceeds 2 K over much of North Africa, Eurasia, and North America, it exceeds 4 K over the interior of North America and Asia, and it exceeds 6 K in central Asia. The change is statistically significant at or above the 5% level in most of the regions enclosed by the 2 K contour. The general pattern of the temperature change (9 kyr BP minus Present) is very similar for the CCM and the LRM (compare Fig. 1a with Fig. 1a of ref. 6) but the magnitude of the change is somewhat larger for the CCM.

Averaged over all Northern Hemisphere land grid points, the increase of surface temperature is 2 K (see Table 1), compared to 1.2 K for the LRM (6). The latitudinal-average difference in land surface temperature (Fig. 2a) reaches almost 4 K around 50°N and exceeds 2 K between 35°N and 60°N; the land is warmer at 9 kyr BP than at present at almost all latitudes north of 30°S (compare with Fig. 2a of ref. 6).

In response to the increased heating and the higher temperature of the land surface, relative to the surrounding ocean which was kept at the present temperature, there is a major redistribution of atmospheric mass. Averaged over all Northern Hemisphere land grid points, the sea-level pressure falls 2.7 mb (equivalent to 5σ and therefore significant at above the 5% level). The decrease in sea-level pressure exceeds 4 mb over much of North Africa and Eurasia and exceeds 8 mb in centers over Central Asia (Fig. 3a). North America is a relatively small continent compared to North Africa-Eurasia, and therefore the region where the pressure decrease reaches 4 mb is small.

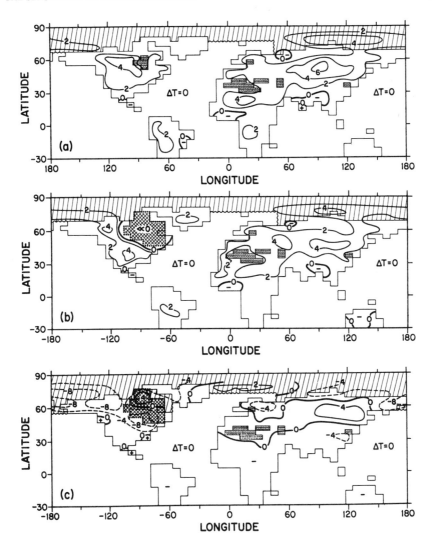

Figure 1 Simulated land surface temperature (K) differences (9 kyr BP minus present) for a) an experiment with solar radiation for July 9 kyr BP, present July ocean surface temperature, and no North American ice sheet ; b) same as (a) except with North American ice-sheet ; c) same as (b) except for January. Temperature difference is prescribed to be zero at ocean gridpoints and at interior sea gridpoints (wavey lines) but not over sea ice (hatched). The area covered by the North American ice sheet is indicated with cross-hatching. Results of the simulation for the region south of 30° s are not shown because the number of land gridpoints is small.

Table 1

Simulated surface temperature (a), precipitation (b), and precipitation-minus-evaporation (c); time averages are July and January; space averages are Northern Hemisphere Land (NH, L), Southern Hemisphere Land (SH, L), Global Average of Land and Ocean (G, L & O), Land region of North Africa [30°N to equator] and the Middle East and Asia [east of 30°E and south of 40°N] (Monsoon). Values are for Modern (M), Standard Deviation of Modern Control (σ), 9000 yr B.P. (9K), Difference (Δ), and statistical significance level (SIG, in percent) of the ratio of the Difference (Δ) to the standard deviation (σ).

a)

TIME AVERAGE	SPACE AVERAGE	control		with ice sheet			without ice sheet		
		M	σ	9K	Δ	SIG	9K	Δ	SIG
July	NH,L	20.7	0.19	22.5	1.8	>1	22.8	2.0	>1
	SH,L	-4.4	0.27	-3.4	1.0	>5	-3.3	1.1	>5
	G,L & O	15.8	0.07	16.2	0.4	>1	16.3	0.5	>1
	MONSOON	24.9	0.14	26.2	1.3	>1	26.3	1.4	>1
Jan	NH,L	-10.8	0.57	-12.3	-1.5	—			
	SH,L	13.2	0.18	11.6	-1.6	>1			
	G,L & O	11	0.10	10.4	-0.6	>1			
	MONSOON	7	0.29	5.3	-1.7	>1			

SURFACE TEMPERATURE (°C)

b)

TIME AVERAGE	SPACE AVERAGE	PRECIPITATION (mm/day)								
		control		with ice sheet			without ice sheet			
		M	σ	9K	Δ	SIG	9K	Δ	SIG	
July	NH,L	4.1	0.10	4.9	0.8	>1	4.9	0.8	>1	
	G,L & O	3.9	0.03	3.9	0	-	3.9	0	-	
	MONSOON	5.5	0.20	6.8	1.3	>1	6.6	1.1	>1	
Jan	NH,L	1.6	0.03	1.7	0.1	-				
	G,L & O	3.6	0.02	3.6	0	-				
	MONSOON	1.6	0.17	1.7	0.1	-				

c)

TIME AVERAGE	SPACE AVERAGE	PRECIPITATION-EVAPORATION (mm/day)			
		control	with ice sheet		
		M	9K	Δ	
July	NH,L	0.28	0.27	-0.01	
	G,L & O	-0.50	-0.50	0	
	MONSOON	1.06	1.26	0.20	
Jan	NH,L	0.37	0.49	0.12	
	G,L & O	-0.53	-0.53	0	
	MONSOON	-0.47	-0.17	0.30	

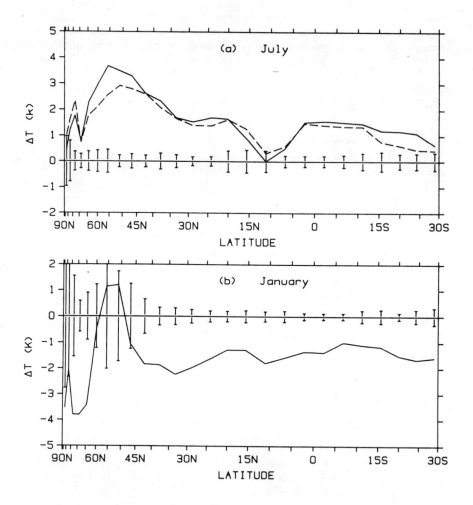

Figure 2 Latitudinal average of simulated land surface temperature (K) difference (9 kyr BP minus present), as function of sine of latitude. Positive values indicate warmer land surface temperature at 9 kyr BP. Vertical bars indicate one standard deviation about the modern control simulation : a) July, b) January. Plots are terminated south of 30°S because the number of land grid points is small. Temperature difference for case with North American ice sheet is indicated with dashed lines in (a).

In contrast to the decreased pressure over land, the pressure is increased over the Indian Ocean and much of the Pacific Ocean. This altered pressure gradient between land and ocean is associated with increased flow of air from ocean to land at low levels and leads to a general strengthening of the Northern Hemisphere summer monsoon circulation at 9 kyr BP. As an indication of the increased southerly flow in low latitudes over the eastern portion of the Northern Hemisphere where the interhemispheric land-ocean contrast is largest, the meridional wind component averaged for 0-180°E and 0-30°N is 1.6 m/s at 9 kyr BP compared to 1.25 m/s for the modern control - a 28% increase.

The precipitation rate over the Northern Hemisphere land is increased by 0.8 mm/day or 20% (Table 1), compared to an increase of 0.5 mm/day (13%) for JJA with the LRM. Over the North African-South Asia sector (identified as the monsoon region, Table 1), the precipitation increase is 1.1 mm/day (20%) compared to 1.3 mm/day (26%) with the LRM. The increased precipitation over the land is compensated by decreased precipitation over the ocean, because the global-average precipitation is unchanged (Table 1).

July (with ice sheet)

The primary impact of the North American ice sheet on surface temperatures is in North America itself (Fig. 1b). At the surface of the ice sheet the temperature for July 9 kyr BP is as much as 10-20 K lower than at the corresponding land surface grid points with no ice sheet. To the west and south of the ice sheet, the temperature increase of 2-4 K for 9 kyr BP is almost identical to that simulated with changed solar radiation alone (Fig. 1a). This result is consistent with geologic evidence that the early Holocene warming occurred much earlier in Central and Western North America - west of the melting ice sheet - than in Labrador and other parts of Eastern North America to the east of the ice sheet (Wright, personal communication). There is a downstream diminution of the maximum temperature increase that resulted from the solar radiation change alone : the maximum temperature increase is 4 K over Asia (Fig. 1b) instead of 6 K (Fig. 1a). Otherwise the spatial pattern of temperature change is very similar, and the latitudinal-average change is similar except between 45-60°N, approximately at the latitudes of the ice sheet (Fig. 2a). Averaged over all Northern Hemisphere land grid points, the increase of temperature is 1.8 K (Table 1).

The July monsoon circulation of 9 kyr BP is intensified (compared to modern) in much the same manner as the case with no ice sheet (compare Fig. 3b with Fig. 3a). Averaged over all

Northern Hemisphere land grid points, the sea-level pressure falls 2.7 mb (equivalent to 5σ). The southerly flow at low latitudes over the eastern portion of the Northern Hemisphere (0-30°N and 0-180°E) is 1.5 m/s, i.e., almost the same magnitude as noted in the 9 kyr BP experiment with no ice sheet. The area-average precipitation over land (Table 1) is nearly identical to the case without the ice sheet.

The latitudinal-average precipitation over land at 9 kyr BP exceeds the precipitation for the control simulation by several standard deviations between 5°N and 30°N (Fig. 4a).

January (with ice sheet)

The decreased solar radiation of January 9 kyr BP causes lower surface temperatures over the continents. Averaged over all Northern Hemisphere land grid points, the decrease is -1.5 K. This decrease is almost identical in magnitude (but opposite in sign) to the July increase (Table 1). The decrease is almost twice that found with the LRM. The colder conditions are simulated at all latitudes except high latitudes of the Northern Hemisphere where wintertime variability is large (Fig. 2b). There are decreased land surface temperatures in the tropics and the subtropics, and at most locations in middle and high latitudes (Fig. 1c). Because the model's natural variability is much larger (in January) in middle and high latitudes of the Northern Hemisphere than in the tropics (Fig. 2b), the relatively small decreases of temperature throughout much of the tropics and subtropics are significant at the 5% level, whereas the 4 K temperature increase in middle latitudes over Asia is not statistically significant.

In response to the decreased heating and lower temperature of the land surface, relative to the surrounding ocean, there is a general rise of pressure over the Northern Hemisphere land amounting to 1.2 mb (Fig. 3c) and an increased northerly component to the low-level outflow from South Asia. Averaged over the same longitude (0-180°E) and latitude band (0-30°N) as before, the northerly flow component is 3.2 m/s at 9 kyr BP compared to 3 m/s in the modern simulation. Immediatly downstream of the North American ice-sheet, sea-level pressure is decreased by as much as 12 mb at 9 kyr BP. This feature is statistically significant at the 5% level.

Changes in land precipitation over the Northern Hemisphere are small (Table 1); an increase of precipitation-minus-evaporation (Table 1) is therefore apparently due to decreased evaporation, but it may not be statistically significant. A feature of potential interest is the increased precipitation over the equatorial and Southern Hemisphere ocean (Fig. 4b ;

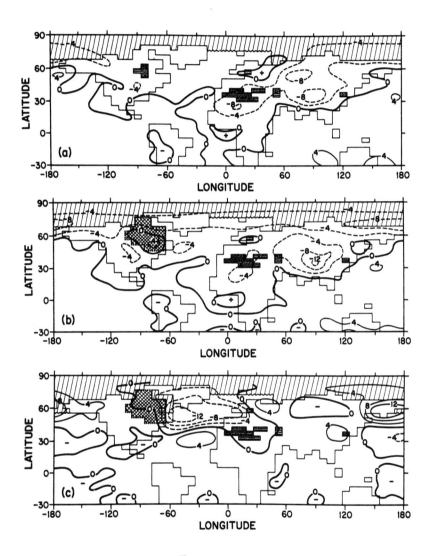

Figure 3   Simulated sea-level pressure (mb) differences (9 kyr BP minus Present) for : a) an experiment with solar radiation for July 9 kyr BP, present July ocean surface temperature, and no North American ice-sheet ; b) same as (a), except with North American ice-sheet ; c) same as (b), except for January.

0-10°S). Examination of the global map of simulated precipitation shows that this increase comes primarily from the South Indian Ocean; it appears to be related to increased low-level flow convergence that is in turn caused by the increased low-level northerly flow of the intensified Northern Hemisphere winter monsoon circulation. If this particular feature is confirmed in subsequent experiments and found to be consistent with geologic evidence for increased early Holocene rainfall south of the equator, then it would provide an example of a Southern Hemisphere climatic change that is "forced" primarily by the thermal response of the N. Hemisphere land surface to the altered solar radiation. In other words, the 9 kyr BP Southern Hemisphere radiation regime of January (southern summer) favors lower temperature and decreased monsoonal rains, but this local radiative response appears in the model to be of less importance than the circulation response to the stronger Northern Hemisphere winter monsoon.

Annual average and range (with ice sheet)

In previous experiments with the LRM, the simulation of a full annual cycle permitted the calculation of annual-average amounts of precipitation and precipitation minus evaporation. Annual-average values are of course essential for detailed analysis of changes in the hydrologic budget. Full seasonal cycle experiments are in progress with the CCM, but, pending their completion, a rough estimate of annual conditions is approximated from the average of July and January conditions. Averaged for all Northern Hemisphere land, the results are as follows : increased precipitation of 0.4 mm/day (10%) or an estimated annual increase of about 150 mm ; increased precipitation-minus-evaporation of 0.06 mm/day or an estimated annual increase of about 20 mm. For North Africa-South Asia (the monsoon region designated in Table 1) the corresponding statistics are : increased precipitation of 0.7 mm/day (20%) or about 250 mm annually; increased precipitation-minus-evaporation of 0.25 mm/day or about 90 mm annually.

In response to the increased amplitude of the seasonal cycle of solar radiation in the Northern Hemisphere, there is a corresponding increase in the annual range of certain climate variables. At 9 kyr BP the annual range of Northern Hemisphere land temperatures (July versus January) was 34.8 K, compared to 31.5 K at present - an increase of 3.3 K (Table 1). In this example, the estimated change of annual mean temperature was small but the seasonality increased appreciably.

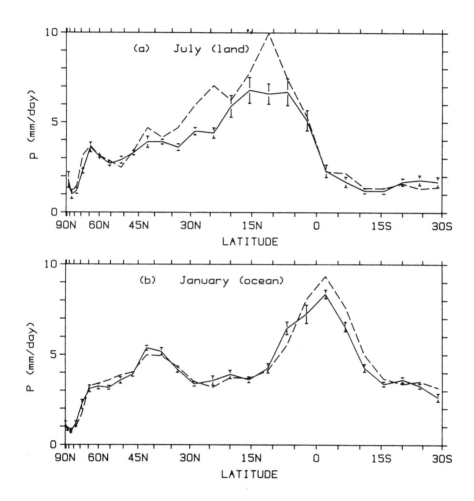

Figure 4 Latitudinal average of simulated precipitation (mm/day) as a function of sine of latitude : a) July--land precipitation; the control simulation (solid line) and the experiment with 9 kyr BP solar radiation and with the North American ice sheet (dashed line) ; b) January--ocean precipitation ; the control simulation (solid line) and the experiment with 9 kyr BP solar radiation and with the North American ice-sheet (dashed line). Vertical bars indicate one standard deviation about the modern control simulation.

RESULTS - RADIATION/CLOUD CLIMATOLOGY

The diagnostics available with the CCM simulations permit an analysis of changes in the radiation, cloudiness, and surface energy budget (Table 2). For the Northern Hemisphere land, the imposed increase in solar radiation for July at the top of the atmosphere was 39 W/m^2. It is of interest to compare that increase with changes at the surface. The increase in net solar radiation at the surface was 20 W/m^2; i.e., more than half of the increased shortwave radiation at the top of the atmosphere was absorbed at the surface. With the warmer land surface and the associated warming and increased moisture content of the atmosphere, the downward-directed longwave radiation increased and therefore the net longwave radiation loss was decreased by 2 W/m^2, producing a total increase in net radiation of 22 W/m^2 over the land. Of the increased net radiation, 21 W/m^2 was expended through increased evaporation. The increased precipitation, expressed in energy terms, was also 21 W/m^2 approximately.

Table 2 Simulated radiation, surface energy and hydrologic budget, and cloudiness changes for the Northern Hemisphere: land and ocean combined (L+O), land only (L), ocean only (O). The July and January differences, 9 kyr BP minus Modern (Δ) are:

Δ SW, top   change in shortwave radiation at top of atmosphere
Δ SW, sfc   change in shortwave radiation at surface
Δ LW, sfc   change in longwave radiation at surface
Δ R, sfc   change in net radiation at surface
Δ E   change in evaporation
Δ P   change in precipitation (both in mm/day and W/m^2)
Δ CLDS   change in total cloud cover (in percent)

	JULY			JAN		
	L+O	L	O	L+O	L	O
ΔSW, top (W/m^2)	37	39	35	-19	-16	-22
ΔSW, sfc (W/m^2)	20	20	20	-12	-10	-13
ΔLW, sfc (W/m^2)	-3	-2	-3	1	-1	2
ΔR, sfc (W/m^2)	23	22	23	-13	-9	-15
ΔE (W/m^2)	1	21	-15	0	-2	1
ΔP (W/m^2)	0.6	20.8	-11.5	0	-1.2	0.6
ΔP (mm/day)	0.02	0.75	-0.41	-0.01	0.07	-0.05
ΔCLDS (percent)	2	-2	3	0	2	-2

Changes in cloud cover occurred at several levels; in general, there was an increased amount of high cloud and decreased low cloud over the land at 9 kyr BP (compared to Present). There was increased low cloud and decreased high cloud over the ocean. These cloudiness changes were consistent with increased rising motion and decreased stability over the land and increased sinking motion and increased stability over the ocean. Because of the vertical compensation of cloud cover changes over both land and ocean, the change in total cloud cover was generally small (2-3%).

COMPARISON WITH EARLIER TIMES

The seasonal (monsoonal) circulations of the present are reduced in amplitude compared to 9 kyr BP. Many examples of similar changes in seasonality exist in the orbital records. Alternations between higher- and lower- amplitude seasonal radiation cycles have time scales that are controled by precession of the vernal equinox with respect to the time of perihelion ($\sim$ 20 kyr cycle), eccentricity ($\sim$ 100 kyr cycle) and, at high and middle latitudes especially, by the axial tilt ($\sim$ 40 kyr cycle). For example, the period from 124 kyr BP to 115 kyr BP, near the end of the previous interglacial, marked a change to decreased seasonality not unlike the change from 9 kyr BP to present (Table 3). The seasonal change of solar radiation was exaggerated around 120 kyr BP because the eccentri-

Table 3 Orbital parameter values (13), and Northern Hemispheric radiation values for various times. "July" and "Jan" for 9, 115, and 124 kyr refer to the appropriate celestial longitudes for comparison with modern July and January (see text).

TIME (kyr BP)	Eccentricity	Tilt	Longitude of Perihelion	Northern Hemisphere Radiation (W/m^2)	
				JULY	JAN
0	0.016724	23.44	102.04	459	231
9	0.019264	24.24	311.22	496	212
115	0.041407	22.41	111.02	430	251
124	0.040287	23.70	323.29	511	207

city was larger then than it is now. A climatic transition from relatively strong to relatively weak Northern Hemisphere monsoon (seasonal) circulations should have been apparent (see also (17)).

An experiment with the CCM for orbital conditions set at 115 kyr BP (being analyzed in conjunction with McIntyre and Ruddiman and Otto-Bliesner) shows decreased July temperature (compared to present) and increased January temperature for Northern Hemisphere land. July precipitation over the monsoon lands is decreased at 115 kyr BP compared to present. The simulated seasonal climatic changes for 115 kyr BP (compared to present) were somewhat similar in pattern to the change "Modern-9 kyr BP)", obtained by reversing the signs of all results reported here. The magnitude of the seasonal change was larger because the eccentricity was larger.

CONCLUSIONS

The results of the experiments for 9 kyr BP with the CCM, a model of higher spatial resolution and improved physical parameterization, broadly confirm the previously reported results (5,6). Changes of the seasonal cycle of solar radiation associated with orbital parameter variations, in combination with the differential thermal response of land and oceans, are likely to have been at least partly responsible for the changes of the climate of the early Holocene. There was a strengthened Northern Hemisphere summer monsoon over the African-Eurasian land mass at 9 kyr BP and also a strengthened Northern Hemisphere winter monsoon. The seasonal range of Northern Hemisphere land temperature was increased. In addition, the seasonal changes of solar radiation over the fast-responding Northern Hemisphere land mass appear to have caused climatic changes south of the equator, over the ocean hemisphere, as well. Another result that is apparent both in previous experiments with the LRM (5,6), in more recent work by North et al. (18), and in the CCM experiments is that the magnitude of the change in land surface temperature and monsoon circulation depends upon the size and latitude of the continent.

Just as the seasonal (monsoonal) contrasts are weaker now than at 9 kyr BP, a somewhat similar decrease in seasonality must have occurred toward the end of the previous interglacial around 120 kyr BP. Thus, changes in the seasonality of climate, or, in other words, changes in the monsoonal characteristics of climate, have occurred many times in climatic history.

ACKNOWLEDGMENTS

Research grants to the University of Wisconsin-Madison from the National Science Foundation's Climate Dynamics Program (ATM-7926039 and ATM-811455 grants) supported this work. The computations were made at the National Center for Atmospheric Research (NCAR), which is sponsored by the National Science Foundation, with a computing grant from the NCAR Computing Facility (35381017). The authors thank W. Washington, M. Blackmon, V. Ramanathan, and B. Boville (all of NCAR) for help and advice in the use of CCM, and M. Woodworth for preparing the manuscript. J.E.K. acknowledges with thanks the opportunity to discuss this work at NCAR as a Summer Visitor to the AAP Climate Section, August 1982.

REFERENCES

1. Milankovitch, M. : 1941, "Canon of Insolation and the Ice-Age Problem. K. Serb. Akad. Geogr.", Spec. Publ. No 132, 484pp. (Translated by Israel Program for Scientific Translations, Jerusalem, 1969, U.S. Dept. of Commerce).
2. Hays, J.D., Imbrie, J., and Shackleton, N.J. : 1976, Science 194, pp. 1121-1132.
3. Imbrie, J., and Imbrie, K.P. : 1979, "Ice Ages : Solving the Mystery", Enslow Publishers, 224 pp.
4. Suarez, M.J., and Held, I.M. : 1976, Nature 263, pp. 46-47.
5. Kutzbach, J.E. : 1981, Science 214, pp. 59-61.
6. Kutzbach, J.E., and Otto-Bliesner, B.L. : 1982, J. Atmos. Sci. 39(6), pp. 1177-1188.
7. Street, F.A., and Grove, A.T. : 1979, Quat. Res. 12, pp. 83- 118.
8. Prell, W.L. : 1978, in : "Evolution des Atmosphères Planétaires et Climatologie de la Terre", Centre National d'Etudes Spatiales (France), pp. 149-156.
9. Kutzbach, J.E. : 1983, The Changing Pulse of the Monsoon. To appear in "Monsoons", Fein and Stephens (Eds), Wiley.
10. Pitcher, E.J., Malone, R.C., Ramanathan, V., Blackmon, M.L., Puri, K., and Bourke, W. : 1983, J. Atmos. Sci. 40, pp. 580-604.
11. Otto-Bliesner, B.L., Branstator, G.W., and Houghton, D.D. : 1982, J. Atmos. Sci. 39, pp. 929-948.
12. Ramanathan, V., Pitcher, E.J., Malone, R.C., and Blackmon, M.L. : 1983, J. Atmos. Sci. 40, pp. 605-630.
13. Berger, A.L. : 1978, J. Atmos. Sci., 35(12), pp. 2362-2367
14. Chervin, R.M. and Schneider, S.H. : 1976, J. Atmos. Sci. 33, pp. 405-412.
15. Ruddiman, W.F., and McIntyre, A. : 1981, Science 212, pp. 617-627.

16. Denton, G.H., and Hughes, T.J. (Eds) : 1981, "The Last Great Ice Sheets", Wiley, New York, 484pp., 28 folded maps.
17. Royer, J.F., Deque, M., and Pestiaux, P. : 1984, in : "Milankovitch and Climate", A. Berger, J. Imbrie, J. Hays, G. Kukla, B. Saltzman (Eds), Reidel Publ. Company, Holland. This volume , p. 733.
18. North, G.R., Mengel, J.G., and Short, D.A. : 1983 : J. Geophys. Res. (submitted)
19. Berger, A.L. : 1979, Nuovo Cimento 2C(1), pp. 63-87

PART IV

CLIMATIC VARIATIONS AT ASTRONOMICAL FREQUENCIES

SUMMARY, CONCLUSIONS AND RECOMMENDATIONS

0. INTRODUCTION

The purpose of the workshop "Climatic Variations at the Astronomical Frequencies" was to stress all the important points that were discussed during the symposium "Milankovitch and Climate". This is why about 30 scientists were asked to present critically their comments by summarizing, for each topics, what were the commonly admitted views before the meeting, what was presented new and worthwhile during this meeting and what are their recommendations for future works. Their final conclusions are reported hereunder and we have assumed that, because of their importance, they will be of great help to the reader.

For the sake of uniformity, the order of the different topics is the same as for the symposium. The principal rapporteurs and contributors are identified at the beginning of each chapter.

1. ORBITAL ELEMENTS AND INSOLATION

Rapporteurs : A.Deprit, P.Bretagnon and A. Berger

1.1. Astronomical solution over the Quaternary period.

Geology brings to astronomy a problem that is entirely new to celestial mechanics. Astronomers never expected that there would be a justified demand for a theory of the solar system to be valid for at least 100 Myr. The elementary analysis of the long periodicities borrowed by Milankovich (1920) from Laplace (1798) and Leverrier (1855) was meant, in their authors' intention, only to confirm stability in the solar system over a few thousand years.

Schematically, the solar system is handled as a set of 10 mass points whose behavior in the field of their mutual gravitational attractions is described by a Hamiltonian in 60 phase variables. In fact the Hamiltonian is usually viewed as the sum $H = S + L$ of a component S responsible for the short period effects, and of a component L, accounting for long periodicities and secular variations.

With a few though notable exceptions, planetaries theories have been essentially directed toward predicting accurate positions over the span of historical and contemporary observations. The general technique has been to apply to H a sequence of transformations (ϕ_1, ϕ_2, ...ϕ_n) so that, at every stage of the induction, the successive short period components $\phi_n(S_{n-1})$ yield progressively their functional importance to the residual components $\phi_n(L_{n-1})$ of long periodicities. A strict application of averaging algorithms in their full generality would lead to literal calculations of truly Herculean proportions. One way of alleviating the task has been to replace sine and cosine functions of long period angles (e.g. the arguments of perigees) by their series in time. For this reason, a theory of the earth's orbit built to serve a bureau of ephemerides is unfit for geological backward extrapolations in time.

Various stratagems have been proposed to overcome this constitutional weakness. One way was to assume that the long term variations of the elements of the earth's orbit may be written as trigonometical expansions over the Quaternary period (Milankovich, 1920; Brouwer, 1965 and Van Woerkom, 1965). If this is true, it is then possible to evaluate the influence of additional terms kept in the disturbing functions and to deduce the improvements upon the solution that will result from such higher order expansion of the solution (Berger, 1977). During this meeting, Bretagnon presented a new long period solution by introducing the relativistic perturbations and the perturbations due to the moon, by improving the integration constants and by adjusting the long period terms on the mean elements of secular variation theories. Following this improvement over his 1974 solution, he concluded that the elements of the orbit are merely defined within some per cent after one million years. Berger has then compared seven such different long period solutions where successive improvements were introduced. Using four different time spans, 800 000 years long, over the last 3.2 Myr, he concluded for the accuracy in time of the related insolation values that improvements are necessary for periods further back than 1.5 Myr BP. However, about the stability of the frequencies, the fundamental periods (around 40, 23, 19 kyr) do not deteriorate with time over the last 5 Myr but their relative importance, for each insolation and for the astronomical parameters, is a function of the period considered.

1.2. Pre-Quaternary values of the astronomical parameters

One would wish now that explorations in the direction undertaken recently by Duriez (1978, 1979) be extended in scope and accuracy. Their results would be adequate over time spans of the order of several million years. However, beyond that point, planetary theories in the service of geology must be completely overhauled.

The traditional viewpoint must be totally reversed: the long period segment L must be given precedence over the short period component S. More precisely, the solar system should be modelled as a coupling of non-linear oscillators (two per planet) with basic periods of the order of 10 kyr. Either one will invent transformations, both concise and quickly convergent, to crush in three to four steps the short period components, or one will discover a technique to reduce the monstrous component S to the surrogate role of providing only the long period terms resulting from resonances as they emanate progressively from the long periodicities. For example, during the symposium, M. Buys and M. Ghil presented the computation of orbital changes as that of free, coupled, nonlinear gravitational oscillators : the planets around the Sun. They pointed out limitations in the accuracy of such calculations, especially the rapid deterioration of phase information, and the slow modification over millions of years of the "fast" periods: 22 kyr and 100 kyr.

Geologists will realize that the new planetary theory on a scale of 100 Myr will take several years to come to maturity. Success depends on imagining unorthodox Lie transformations with unusual purposes, and in experimenting with them. It depends also in a critical manner on new software techniques, probably to be implemented on LISP machines, especially if they require processing elliptic functions in a literal way.

Soon thereafter will come a time when geology will provide astronomers with estimates of periodicities in significant disagreement with the calculated values. This would be the most interesting moment for celestial mechanics. Indeed, at that point, astronomers will be allowed to leave their role as accountants of the gravitational budgets in the solar system, and to discuss with geologists which paleastronomical effects should be added to the gravitational model in order to reduce the disagreements.

1.3. Insolation parameters

As the climate system is thermally driven by solar insolation there is a real interest to check whether or not there are some relationships between insolation parameters and the climate at the global scale. The computation, accuracy and spectrum of the different kinds of insolation, which are supposed to be used for modelling climate or for simulating climatic variations, were carefully reviewed by Berger during the symposium. His conclusions are summarized in Table 1.

Other works about insolation parameters were also presented, namely those dealing with latitudinal gradients of insolation (cfr the chapter on Conceptual Models). Fourier representation of orbitally induced perturbations in seasonal insolation was computed by Taylor with the aim of improving our intuitive sense of how each of the orbital parameter affects insolation.

1.4. References

Berger A.L.: 1977, Celestial Mechanics 15, pp.53-74.
Bretagnon P.: 1974, Astronomy and Astrophysics, 30, pp.141-154
Brouwer D.: 1963, in :"Climatic Change", H. Sharpley, Harvard University Press, Cambridge, pp. 159-164
Duriez L.:1978, in: " Dynamics of Planets and Satellites and Theories of their Motion". V.G. Szebzhely (ed.), Astr. and Sp. Science Lib., 72, pp. 15-32
Duriez L.: 1979, Approche d'une Théorie Générale Planétaire en Variables Elliptiques Héliocentriques. Thèse de Docteur ès Sciences. Lille.
Laplace P.S.: 1978, Traité de Mécanique Céleste. Paris
Leverrier U.J.J.: 1855, Recherches Astronomiques, Annales de l'Observatoire Impérial de Paris.
Milankovitch M.: 1920, Théorie Mathématique des Phénomènes Thermiques produits par la Radiation Solaire. Académie Yougoslave des Sciences et des Arts de Zagreb, Gauthier--Villars, Paris
Van Woerkom A.J.J.: 1963, in : "Climatic Change", H.Shapley (ed.), Harvard University Press, Cambridge. pp. 147-158.

Table 1 - Insolations as a function of astronomical parameters (++ means stronger dependancy).

	ε	$e \sin \tilde{\omega}$
Mid-month insolation at equinox		+
at solstice	+	++
Half-year astronomical seasons		
- total insolation	+	
- length		+
- mean in polar latitudes	++	+
in equatorial latitudes		+
Caloric seasons polar latitudes	+	
equatorial latitudes		+
Meteorological seasons (astronomical definition		
- total insolation	+	
- length		+
Meteorological seasons (monthly mean)	+	++

2. PRE-PLEISTOCENE EVIDENCE OF ORBITAL FORCING

Rapporteurs : P.L. de Boer and N. Shackleton.

2.1. Historical background

The first point that we noticed with reference to the Pre-pleistocene record of Earth-orbit effects, is that although we associate Milankovitch's name with the Quaternary ice ages, the earliest evidencies in the geological record for orbital cycles were found in pre-Pleistocene sequences. It is only in the past twenty years that it has become possible to obtain Pleistocene records from the deep sea that provide timeseries suitable for detecting orbital effects. On the continents the effects of orbital forcing were mostly far too dramatic to leave records for analysis.

The good stratigraphic control and the relatively easy accessibility of Pleistocene sediments has made that the development of theories about astronomical forcing of climates has mainly concentrated on these during the last decades. For pre-Pleistocene sediments the recognition of astronomical periodicities has mainly been an additional result of geological studies.

The name of G.K. Gilbert (1895) is particularly important as he noted the prevalence of cycles of probably orbital origin and proposed their use for the measurement of geologic time. Another key worker was W.H. Bradley (1929), who demonstrated the record of the precessional cycle in the Eocene Green River Formation with a calibration based on varva counts. More recently, W. Schwarzacher (1954), A.G. Fischer (1964), F.B. Houten (1964), and many others have worked on cyclic sedimentation and its probable relation to orbital changes. On the whole, however, their work circulates among geologists interested in the phenomenon of cyclic sedimentation, rather than amongst palaeoclimatologists or climate modellers. This meeting may represent the first occasion on which so much of the pre-Pleistocene data has been shown to an audience interested in the underlying forcing rather than in the deposits.

2.2. Results shown at the meeting

The oldest cycles, reported by Anderson from the Castille Formation in Texas, consist of fine layers that he convincingly argued to be annual; the sequence is only about 200 kyr long, but the duration is known very accurately and the data is probably better than anything else in the geological column for spectral analysis over the whole bandwith from annual to Milankovitch frequencies; unfortunately, the record length is not sufficient to provide unambiguous evidence for the effect of eccentricity changes. Olsen's data from the Triassic in the New Jersey region, which is an extension of van Houten's work, also has the benefit of varve calibration and was very convincing.

Fischer and de Boer both demonstrated the Cretaceous data from Italian sections with the aid of dramatic pictures of the rocks as they strike the eye. The "bundling" of beds repeating on about a 20 kyr timescale, into groups of four or five, provides good circumstantial evidence that a precession cycle modified by eccentricity, is responsible. It was however freely admitted that the geological controls on the periodicity are rather weak and the uncertainty on the dominant period could be as much as 50% to judge from the literature age data. However, the similarity of frequency spectra based on the measurement of the carbonate-marl rhythms and on certain elements of orbital parameters, which at present define climatical changes, strongly suggests a causal relationship between orbital influences and the Cretaceous pelagic sedimentary cycles shown.

Closer to the present day, three presentations dealt with the one or two millon years prior to the million-year past which has been the main focus of Milankovitch studies in the last few years: Shackleton's and Schnitker's dealt with the oceans, and Hooghiemstra's with pollen data from Columbia. Here the chief interest at present is in observing the gradual change in the mode of response of the Earth's climate system to orbital forcing. The earlier parts of these records showed less evidence for the eccentricity cycle, and hence an impression of greater high frequency variation, before about 2.5 million years ago. This indeed suggests the pronounced 100 kyr signal during the Pleistocene to have been strongly ruled by icecap dynamics.

With regard to modelling the orbital effects for a preglacial Earth, Arthur's talk was particularly significant, in that he emphasized the important role that diagenetic alteration to the sediments after deposition may have in amplifying or obliterating the geological signal which may originally have been very subtle. This was also evident in de Boer's paper in that weathering at the outcrop revealed extremely clear cyclic sequences which were far less impressive in chemical analytical data. Arthur's presentation also made it clear that much of the deposits available for study are probably totally unsuitable for most types of time series analysis, so that the variety of unconventional mathematical techniques used should not be regarded as naive so much as ingenious.

Paleomagnetic data from the Cretaceous of VandenBerg suggest that the magnetic field of the Earth also responds to orbital forcing. Despite the extensive paleomagnetic research that has been done during the last decades, this would be the first time that such is recognized.

2.3. Recommandations

The fossil sedimentary record offers a 1 to 1 scale model in which all processes involved have worked out from the very start to the very end. Cyclic sediments such as discussed during the meeting, are present on many places and in many stratigraphic intervals, and they may provide much information about orbital movements in the past. For the subrecent, astronomical parameters are used as a standard to compare atmospherical, oceanographic and sedimentological processes, but for the geological past the sedimentary record reversibly can be used to establish the former behaviour of the Sun-Earth-Moon system.

It is clear that deposits in which annual layers may be counted to provide a time control, are of supreme importance and amply justify the amount of work that their study entails; deposits such as in the Castille formation are extremely precious geological archives, while deposits such as the Miocene-Pliocene Monterey diatom-varved material should certainly be analysed.

Arthur's presentation gave clues to possible means of expressing sequences such as those described by Fischer on a scale that is more closely related to time than is observed thickness; future work on these type of deposits will be of even greater importance if this can be achieved. In this respect, detailed paleomagnetic studies might, in the view of the apparent relation between orbital parameters and the Earth's magnetic field shown during the meeting, offer an excellent control of relative, and in a later stage possibly also of absolute time.

Additionally it will be increasingly important to identify the range of different controls that can be demonstrated to generate detectable variations in geological deposits resulting from orbital forcing. It has already become clear that much of the material, displaying apparent precessional forcing, is from low paleolatitudes, where modelling has already suggested that a significant response does not depend on ice-sheet feedback.

On the other hand much of the Tertiary Pre-Pleistocene deep sea data points to tilt control and perhaps to deep-water formation at high latitudes as the origin of the signal.

Pollen records provide evidence of vegetational changes that may also be very sensitive monitors of subtle climatic changes, but there is little tradition of mathematical analysis in such data sets. We recommend that the numerical studies, such as the one of Hooghiemstra may generate, should be given the mathematical attention that they deserve.

The period leading up to the Pleistocene glaciations is particularly important because here geological time may mimic the tuning that is put into attempts to model the past million years, and because we have sufficiently good temporal control to monitor the response of many parts of the climatic system. Here, the deep sea seems the repository of the best materials, along with infrequent, but very important, lake basins. Sampling these requires an investment in coring to recover undisturbed sequences through the deposits, and we emphasize the importance not only of the Hydraulic Piston Corer which has brought Glomar Challenger program to the limelight in paleoclimate research, but also other means of recovering the sediment such as the Long Coring Facility that is being developed.

Effort must also be put into recovering long lake sections from those basins that preserve long records. The study of such sections becomes increasingly valuable as more and more studies are performed in collaboration, especially if identical samples can be studied by several workers, so that phase relations between different signals are preserved. Although this has become a traditional approach in the study of deep sea sediments, its important is not confined to such materials.

2.4. References

Bradley, W.H.: 1929, U.S. Geol. Surv. Prof. Pap. 158-E, pp.87-110.
Fisher, A.G.: 1964, Kansas Geol. Survey Bull. 169,pp.107-149
Gilbert, G.K.: 1895, J. Geol. 3, pp.121-127
Schwarzacher, W.:1954, Thermaks Mineral. Petrograph. Mitt. 4,pp.44-54
van Houten, F.B.: 1964, Kansas Geol. Survey Bull. 169, pp.497-531

3. MARINE PLEISTOCENE RECORDS OF CLIMATIC RESPONSE

Rapporteurs : J. Hays, A. McIntyre, A. Bloom, J.C. Duplessy

3.1. Proxy indicators of oceanic response to climate change (PIOR)

The long history of geologists using proxy indicators of oceanic responses to climate change in sediments evolved in the late 1970's to the CLIMAP reconstruction of ocean surface conditions (sea surface temperature) at 18k and 125k (1981 and 1984). Thus, Geological oceanographers and modelers came to the NATO conference with a good picture of the spatial variability of sea temperature estimated from microfossils, at these two times. Since faunal assemblages in sediments can also be related to specific water masses, the CLIMAP data also showed their distribution and by inference, ocean surface circulation.

This conference, however, highlighted (i) the rapid expansion of information on temporal changes in sea surface conditions with Pleistocene sea surface temperature estimates from microfossil data, for the North Atlantic (Ruddiman and Mc Intyre), the equatorial Atlantic (McIntyre and Ruddiman), the North pacific (Pisias and Leinen) and the Indian Ocean upwelling system in the Arabian Sea (Prell), and (ii) their use in making conceptual models of ocean response to climate. Along with surface conditions, time series of proxy indicators of surface circulation change and orbital forcing were discussed. Prell highlighted the use of a microfossil, G. bulloides, as an indicator of circulation processes, specifically, upwelling intensity in the Western Arabian Sea. Pisias and Leinen were able to show that a transitional radiolarian assemblage and biogenic silica in a Northwest Pacific core record the N-S movement of the Kuroshio-Qyashio confluence.

The use of benthonic foraminiferal assemblages as proxy indicators of bottom water conditions and circulaion has been documented before this meeting in the work of Streeter, Schnitker, Corlin, and Lohman, all of whom have mapped the distribution of assemblages with respect to present day bottom water masses and conditions using the transfer function approach. At this meeting Schnitker emphasized the temporal variability of benthonic foraminiferal assemblages in the North Atlantic both during the Pleistocene and the Pliocene.

3.2. Proxy indicators of atmospheric activity (PIAA)

PIAA reveal two general types of paleoclimatic information: first, indications of continental climate, and secondly, qualitative and quantitative indicators of the intensity of atmospheric circulation. These sorts of indicators have been studied for roughly 25 years. Contributions of Rex and Goldberg (1958), Windom (1976), and of Prospero (1981) on dust input to and over the ocean and of Gillette (1981) on mechanisms of dust transport have been particularly valuable. The continental record of climate is vast, for the Pleistocene important contributions have been made by those who study past take levels and areal extent of arid regions. Marginal oceanic basins contain records of continental runoff, presumed to reflect rainfall, either in detrital sediment accumulation rates or, more rarely, sapropel formation and sea-surface salinity changes. The pelagic oceans also have been found to contain a record of continental activity preserved as the variable flux dust to the sea floor. Indicators of the intensity of atmospheric circulation have been found in surface circulation, particularly upwelling indicators (indirect) and in the grain size of the eolian dust (direct).

Presentations at the meeting served to bring the PIAA to the attention of the broader community of climate modellers and to advance the general understanding of the nature and variability of these indicators, especially variability in the frequency domain. Rossignol-Strick associated a 500 kyr Mediterraneous sapropel record with runoff and related that to the intensity of monsoonal circulation (strong seasonal changes) as controlled by fluctuations in obliquity. Prell discussed a 100 kyr upwelling records from the northwestern Indian Ocean which he related to the intensity of monsoonal circulation there. That record contained spectral power at the precessional frequencies of 23 and 19 kyr. North Pacific dust accumulation records spanning Bruhnes presented by Janecek and Rea record variations of source-area climate generally indicating humid glacials and arid interglacials (with some lags). The dust grain-size record of the history of wind intensity, shows fluctuations of ± 20 to 45% with larger fluctuations occuring before 250 kyr ago. Spectral analyses of the grainsize record revealed frequencies of 104, 41, and 23 kyr in the westerlies and of 140 and 40 kyr in the tradewinds (one core each). Both direct (quartz) and indirect (faunal and opal) records of atmospheric circulation were presented by Pisias and Leinen. Their records show spectral power at both obliquity (in indirect indicators) and precessional (quartz) frequencies. A change in atmospheric circulation at 250 kyr is recorded in this data set also. Together the presentations on North Pacific PIAA emphasize the importance of regional variability in atmospheric and oceanic circulation.

The interaction of people and disciplines incurred by this meeting made obvious several important avenues of research critical to developing a full understanding and use of PIAA. To better understand the indicators themselves, we particularly encourage: attempts to collect PIAA data from cores where other paleoclimatic proxy indicators are avalaible; and finding "tags" in the eolians components that would aid in determining continental source area and climate ($\delta^{18}O$ of minerals, pollen, etc.). PIAA should be utilized to determine the climatic (precipitation - evaporation, runoff) history of continental source regions, and changes in the intensity of atmospheric circulation through geologic time. The spectral records generated will change in space (tradewinds, westerlies,southern hemisphere) and time, and a comprehension of these variations will be important in determining the atmospheric response to internal and external (Milankovitch) forcing. Finally, the best ultimate use of PIAA may be to test the output of, and to provide input to computer models of atmospheric circulation and global climate.

3.3. Time scale and oceanic Pleistocene response in the frequency domain

An accurate Pleistocene time scale for deep sea cores is a basic premise to compare variations of the Earth's insolation with those of climate proxy-indicators. Therefore the features common in all isotope records have been identified. This graphic correlation provides exact correlations from core to core and allows the identification of hiatuses, coring disturbances and naturally induced changes in the sedimentation rates. An iterative adjustment between the three available radiometric control points (isotopic transition 5e/6 at 127 kyr; magnetic reversals at 730 kyr and 910 kyr BP) has been used to maximize coherencies between orbital and isotope signals at the three frequencies (41 kyr; 23 kyr; 19 kyr) around which obliquity and precession variance is concentrated. The resulting cross-spectrum has significant coherencies not only at those frequencies but also at 100 kyr, indicating that the continental ice volume is also driven by the eccentricity. The phase spectrum is satisfactorily modeled for the three higher frequencies as a single exponential system but at lower frequencies, observed phases and amplitudes significantly disagree with the model, suggesting a non linear resonance near 100 kyr.

In fact, time-series of discrete components of varying provenance in the oceanic Pleistocene record yield signals with significant frequencies correlative with these primary orbital frequencies of precession and obliquity. The components may be subdivided on the basis of provenance into those from surface water, deep-water and continents. The fact that three such disparate groups record what are presumably orbital frequencies strengthens the hypothesis of a cause-effect relationship.

Surface waters : Ruddiman and McIntyre (1984) demonstrated that the high latitude fluctuations of the northern part of the N. Atlantic subtropical gyre is dominated by the 41 kyr obliquity signal. Cores from the southern part of the subtropical gyre and at the Equator contain the strongest 23 kyr (precession) signal yet recorded, while in these same cores the obliquity is not significant. In the Antarctic, Hays, Morley and Shackleton (using the radiolarian species Cycladophora davisiana) have shown that both Antarctic and Arctic latitude N. Pacific cores contain the 41 kyr and 23 kyr frequencies, while this same species records only the 23 kyr and 19 kyr frequencies in the Equatorial Pacific. Similar results were documented by Pisias and Leinen in North Pacific cores where Radiolarian time series contained 100, 41 and 23 kyr frequencies. In addition, organic opal registered 40 kyr.

Deep water : Schnitker utilizing benthonic foraminifers demonstrated an increase in faunal amplitude fluctuations with time, over the Milankovitch band, as well as a higher frequency signal of unknown origin.

Continental : Janecek and Rea obtained time-series of eolian transported sediments from a Pacific core underlying the Northern Hemisphere mid-latitude westerlies and the equatorial easterlies. Spectral analysis of the former yielded frequencies of 104, 41 and 23 kyr while the latter only contained 140 kyr and 40 kyr (because the sample interval was too coarse). To resolve frequencies in the range of 23 kyr, Pisias and Leinen also obtained eolian time series in their NW Pacific core with a frequency of 23 kyr.

3.4. Phase relationships

Phase relationships between various climatic parameters were studied within specific frequency bands by several investigators. Prior to the meeting, phase relationships between various proxy climatic records had been compared by Hays et al. (1976) who observed that at the 41 kyr period, $\delta^{18}O$ lagged the obliquity by ~ 9 000 years, while at 23 kyr $\delta^{18}O$ lagged precession by about ~ 5 000 yrs. In the Antarctic section, they also noted that the record of estimated summer temperature lead the $\delta^{18}O$ record by about 2 000 years.

Ruddiman and McIntyre (1981) noted that their estimated summer temperature record in the North Atlantic was in phase with the $\delta^{18}O$ record at times of deglaciation but lagged the $\delta^{18}O$ record during times of ice growth.

At this meeting the phasing of various proxy climatic records was compared with orbital signals and with the $\delta^{18}O$ record at global ice volume. The results are summarized above in Section 3.3.

3.5. References

CLIMAP Project Members, A. McIntyre (leader) : 1981, The Last Glacial Maximum. Geol. Soc. America Map and Chart Series 36.
CLIMAP Project members, W.F. Ruddiman (leader), R.M. Cline (Ed.) : 1984, The Last Interglacial Ocean. Quaternary Research (in press).
Gillette, D.A. : 1981, in : "Desert Dust : origin, characteristics, and effect on man", T.L. Péwé (Ed.), pp. 11-26, Special Paper 186, GSA, Boulder, Co.
Prospero, J.M. : 1981, in : "The Oceanic Lithosphere", C. Emiliani (Ed.), pp. 801-874, John Wiley and Sons, N.Y.
Rex, R.W., and Goldberg, E.D. : 1958, Tellus 10, pp. 153-159.
Ruddiman, W.F., and McIntyre, 1984, Ice-Age thermal response and climatic role of the surface Atlantic Ocean 40°N to 63°N, Geol. Soc. America (in press).
Windom, H.L. : 1976, in : "Chemical Oceanography (2nd ed.), vol. 5, pp. 103-135, J.Pl. Riley and G. Skirrow (Eds), New York, Academic Press.

4. PLIO-PLEISTOCENE LAND RECORDS.

Rapporteurs : G. Kukla and P. Aharon

4.1. Preambule

There are only few stratigraphic records on land that are sufficiently well dated, continuous, long enough to cover several periods of orbital perturbations, and containing clear climatic indicators. The paucity of high-quality records on land is however, well compensated by the diversity of proxy climatic indicators that in conjunction offer valuable insights into the nature of Pleistocene climatic changes. Land-based records thus may provide useful information on (i) paleo-sea-levels and temperature of the marginal oceans; (ii) surface air temperatures on land; (iii) shifts in the evaporation/precipitation ratios (E/P) and patterns of vegetational changes; (iv) variations in the wind direction and intensity; (v) transgressive-regressive cycles of continental ice caps and permanent mountain snow fields.

The best dated land-based proxy climate records are: (i) emerged coralreefs and beach-ridges; (ii) pollen-bearing lake beds; (iii) loess-soil sequences; (iv) valley-fills related to the glacial deposits; (v) terminal moraines on tropical highlands; (vi) speleothems.

Only the first two areas were addressed at the symposium. The most recent results are briefly summarized below, and the recommandations for future investigations are outlined.

4.2. Raised coral-reefs

Coral-reef terraces of late Pleistocene age that are preserved on tectonically active coasts (e.g. Barbados, New Guinea) offer unique record that includes (a) radiometric dates of uncristallized corals and molluscs, (b) paleo-sealevel estimates, (c) $^{18}O/^{16}O$ isotopic record that reflects shifts of the global ice volume and of the local sea water temperature; (d) $^{13}C/^{12}C$ record that reflects variations of the carbon cycle in the marginal seas; (e) record of seasonal changes in the temperature and salinity of sea water preserved in the isotope ratios of individual annual growth increments of the coral-reef biota.

Previous work on coral terraces from Barbados confirmed the existence of the ca. 20 kyr precessional cycle in climate by radiometric dating of the reefs to 125, 105 and 85 kyr (respectively (Broecker et al., 1968; mesolella et al., 1969). The $^{18}O/^{16}O$ measurements of corals associated with the reefs also confirmed the validity of paleo-sealevel estimates during the early Wisconsian high sealevel stands (Fairbanks and Matthews, 1978). Subsequent studies of a well subdivided sequence of emerged cora-reef in New Guinea has confirmed the data from Barbados and also identified a series of high sealevel stands during the late Wisconsian at 60, 42 and 28 kyr respectively (Chappell, 1974; Bloom et al., 1974).

More recent $^{18}O/^{16}O$ measurements of giant clams from the New Guinea reefs yield a detailed record of ice volume and temperature changes over the last 10^5 yrs (Aharon, 1980). The timing, frequency and intensity of climatic events recognized in New Guinea closely parallels the insolation intensity shifts at sensitive high latitudes predicted by the astronomical theory of ice ages (Berger, 1978). This implies that a cause and effect relationship exists between the two records. There are still some discrepancies between the predictions of the astronomical theory and the coral-reef data. These are most conspicuous during the time of the last interglacial and further investigations are required. Isotope measurements of coral-reef sequences from New Guinea that span the last 400 kyr, now in progress, may well resolve some of the existing inconsistencies.

4.3. Pollen-bearing lake beds

Three continuous pollen records of the Late Pleistocene were discussed. All are from the temperate zone of the northern hemisphere: Grande Pile in northeastern France (L. Heusser, B. Molfino and G. Woillard); Clear Lake in northwestern U.S. (D.P. Adam) and Biwa Lake in Japan (S. Kanari, N. Fuji and S. Horie). The first two studies spanned only the last interglacial/glacial cycle ($\sim$0-130 kyr) while the third spanned a much longer time period ($\sim$0-400 kyr).

Variations in pollen signature indicating changes of the relative abundance of climatically sensitive taxa and of interpreted paleotemperatures from these areas showed good correspondence. All three records indicate climatic variations in accord with the known Milankovitch periodicities, but quantitative analysis was only possible in the Grande Pile where a reliable stratigraphic framework permitted correlations in both frequency and time domains. This framework includes the detailed ^{14}C control, close correlations with the oceanic oxygen isotope record, and correlation with radiometrically dated Pleistocene pollen sequences elsewhere in Europe.

Spectral analysis revealed a significant effect of the precessional cycle on the pollen frequencies. Statistically significant peaks were found at the 23 kyr periodicity as well as the 10 kyr combination tone of the precessional periodicities (23 kyr, 19 kyr). Although the obliquity and eccentricity cycles were unresolved due to the shortness of the pollen record, the Milankovitch periodicities of 19 kyr, 23 kyr and 41 kyr account for 40-60% of the variance in this proxy climatic data.

A better time-stratigraphic control and longer pollen sequences are of prime importance in the accurate determination of Milankovitch effects on the continental record. More importantly, such studies can provide information on eventual lag response of the climate system to insolation forcing. Studies of the long core raised from the Lake Biwa in Japan promises to provide such data.

4.4. Recommendations

Contributions presented at the Milankovitch and Climate Symposium confirmed the existence of close correlations between terrestrial climates and the orbital perturbations. They also identified problems whose solutions call for improved sampling strategy as well as analytical techniques. Recommendations emerging from those contributions may be summarized along the following lines:

(i) It is imperative to establish (a) a reputable international program to raise continuous long sedimentary records from lake beds in climate-sensitive locations spanning the two hemispheres; (b) an internationally organized multidisciplinary effort to analyze the data and oversee the central storage of the cores (i.e. a land equivalent of the DSDP or IDOP projects)

(ii) The radiometric dating of critical climatostratigraphic horizons has to be improved. These include among others (a) the last interglacial; (b) the Brunhes-Matuyama; (c) major continental glaciations. Needed refine of the time scale can be achieved by a combination of radiocarbon, fission track, K/Ar, thermoluminiscense and tephrochronology techniques.

(iii) The understanding of climatic signifiance of proxy climatic indicators has to be refined.

(iv) The resolution of the isotope measurements in corals should be improved to a degree necessary to unravel seasonal variations in the sea surface parameters.

(v) Correlation of the non-marine, estuarine, marginal marine and deep-sea should be continued emphasizing ties with the major glacial episodes on land.

4.5. References

Aharon, P. : 1980, Stable Isotope Geochemistry of a Late Quaternary Coral Reef Sequence, New Guinea. Doctoral Theses, Australian National University, Canberra.
Berger A.L.: 1978, Quaternary Reseach, 9, pp.139-167
Bloom A.L., Broecker W.S., Chappell J.M.A., Matthews R.K. and Mesolella K.J.: 1974, Quaternary Reseach 4(2), pp.185-205
Broecker W.S., Thurber D.L.,Goddard J., Ku T., Matthews R.K. and Mesolella K.J.: 1968, Science, 159, pp.297-300
Chappell J.M.A.: Nature, 252, pp.199-202
Fairbanks R. and Matthews R.K.: 1978, Quaternary Research 10, pp.181-196
Mesolella K.J., Matthews R.K., Broecker W.S. and Thurber D.L.: 1969, J. of Geology, 77, pp.250-274

5. CORRELATIONS BETWEEN MARINE AND NON-MARINE PLEISTOCENE RECORDS.

Rapporteurs : D. Adam and M. Sarnthein

It was apparent at the conference that almost no continental climatic records are continuous, well-dated, and long enough to permit spectral analysis at Milankovitch frequencies. On the other hand, the few long continental records available (Lake Biwa, Japan, the $3.5*10^6$ year record from Funza, Colombia reported at this meeting, and the presently unpublished 0.9 Myr Macedonian core of Wijmstra) offer a higher temporal resolution than is available in most deep-sea cores. Cyclical records of aeolian dust and pollen in deep-sea cores show that continental environments were affected by Milankovitch cycles, a view confirmed by the few long continental records.

Understanding the interactions between orbital variations and the climatic system will require evaluation of the phase relationships between oceanic and continental records. Whereas most deep-sea cores can be directly correlated with the global oxygen-isotope (ice volume) record, this is not possible for continental records. To avoid circular reasoning, it is necessary to establish direct correlations between continental and marine deposits that are independent of curve-matching.

A number of physical factors and processes affect both sea and land and are potentially available for time series analysis. Study of leads and lags between marine and continental time series will lead to a better understanding of the long-term behavior of the climatic system. Significant factors inferred from proxy data in the stratigraphic record include:

1) wind regimes, such as monsoon tracks, westerlies, trades, the position of the Hadley Cell; (e.g. by pollen flux, dust flux, dunes)
2) land albedo, (e.g. by lake levels, paleo-vegetation, ice distribution)
3) evaporite formation
4) meltwater flux (retreating moraine systems, oxygen isotopes in the marine record, fluvial sediment bodies at the continental margin)
5) sea-ice boundary (e.g. by drop stones, diatom blooms, etc. in marine sediments)
6) the CO_2 balance (e.g. by vegetation record, carbonate deposition).

At the conference, major contributions were presented by :

- Adam, Heusser et al., Hooghiemstra, Hories et al. with regard to long-term section (pollen evidence)
- Aharon, Fairbridge, Peltier concerning the sea-level record
- Herterich et al., Janecek et al., Pisias et al., Rossignol-Strick concerning deap-sea sediment sections with a land record
- Fillon, Hays et al., Ruddiman, Mc Intyre, to the problem of land and sea-ice distribution.

5.1. Recommendations

1) Clear stratigraphic correlations independent of curve-matching are essentially before time series from land and marine sections can be compared in the frequency domain. We recommend further detailed work on methods to achieve such correlations, including tephrochronology, paleomagnetism, radiometric and other absolute dating methods, and study of annually laminated deposits.

2) Sections should be analyzed from a range of continental localities characterized by continuous deposition for at least 200 000 years (i.e., at least two 10^5-year cycles). These sections should be thoroughly studied for multiple climatic time series and carefully dated using the methods mentioned in (1) above. As a long-term goal, longitudinal transects of these sections should be studied, both in the Western and Eastern Hemispheres. Lake sediments and loess sections are particularly promising.

The hydraulic piston corer should be adapted for use in land- and lake-based drilling operations in order to permit the recovery of core sections that can be self-oriented for paleomagnetic declination cycle studies.

3) Sea-level extremes should be studied in detail in relation to both the record of continental coastal environments and the oceanic oxygen-isotope record, and seismic stratigraphy.

Moreover, shoreline deposits often provide material which can be exactly dated. Other detailed studies of interfingering marine and continental deposits along the continental margins may also be productive.

4) Deep-sea sediment sections downwind or downstream from continental sediment sources can provide detailed long-term time series of terrestrial environmental changes of the past 5 to 10 Myr through analysis of both inorganic and organic components. These studies should be intensified in both the Northern and Southern hemipheres, particularly along continental margins.

SUMMARY, CONCLUSIONS AND RECOMMENDATIONS

6. ESTIMATION OF CLIMATE SPECTRA

Rapporteurs : N. Pisias and P. Pestiaux
Contributors : H. Dalfes and K. Herterich

6.1. Present state of knowledge

A fundamental goal of paleoclimatic research is to explain all sources of variance in the Earth's climate - both variance associated with spectral lines and variance found in the spectral continuum. The first step towards this goal is the definition of the spectrum of global climate change over intervals of several hundreds of thousands of years. Since all estimates of climate spectra over this interval come from the geologic record (dominately deep-sea sediment cores) these estimated spectra reflect the problems and assumptions associated with the study of geologic time series as well as the analytic/statistical techniques used.

Most studies analyzing geologic time series have relied on the classical technique of Blackman-Tuckey method where spectra and cross-spectra are calculated from the autocovariance and cross-covariance functions. Some studies have used the fast-Fourier-transform to calculate variance spectra. These methods have been well studied and the statistical properties of the spectral estimates are well understood. These techniques usually produce identical results but the fast-Fourier-transform procedures have an advantage of reduced bias in the spectral estimates whereas the Blackman-Tuckey method has the important advantage of being able to handle time series with missing data values (a common problem with geologic data sets).

When calculating a spectrum, two basic assumptions are made independent of the technique used:
(i) - The time scale of the data series is known and
(ii) - The data set is stationary at least over the interval being studied.

The first assumption is a major problem in the study of paleoclimate records. With the exception of three-ring studies and annually laminated sediment, there are no direct means of measuring time in a geologic section. In studies of deep-sea sediments, samples can be taken at known depths in the sediment column but the ages of each sample are not directly measured. The assumption of constant sediment accumulation rates is often made so that depth in section can be related to time, but in many parts of the world's ocean this assumption has been shown to be false. Thus, much effort has been directed towards the establishment of a highly accurate geologic time scale for the last several hundred thousand years (Hays et al., 1976; Kominz et al., 1979; Morley and Hays, 1980; Johnson, 1982). These workers have used the hypothesis of orbital forcing of climate change to adjust the time scale of geologic records to better "fit" the calculated orbital parameters. This approch reduces the independence of geologic records as verification of Milankovitch hypothesis, however, the high coherence is observed between the proxy-climate data and all orbital parameter strongly argue for the presence of orbital forcing in the climate system (Imbrie et al., this volume).

The degree to which the climate system is non-stationary over thousands or millions of years starts only now to be studied in terms of how the variance spectrum of climate has changed. Pisias and Moore (1981) have shown that over the last 2 millions years significant changes in the spectrum of ice-volume changes have occured. Most of those changes have occured in the long-period components of the climate record. As longer geologic time series are studied, the problem of detecting and defining non-stationary in the climate system becomes much more important.

Presently, it is assumed that spectra of climate proxies reflect directly spectra of actual climate variables. There has been no attempt to understand the degree to which proxi-climate spectra reflect the true spectra of climate. The transfer from the true climate spectrum to the proxi-climate spectrum can be considered as a series of frequency transfer functions. Variability in the global climate system changes many aspects of the Earth's environment. These changes, such as the growth and decay of glacial ice, are recorded in the geologic record. In the case of glacial ice growth, changes in ice-volume are recorded in the isotopic composition of the world's ocean due to isotopic fractionation. Changes in the isotopic composition of the ocean depend on the total volume of glacial ice stored on land, and the degree to which each incremental unit of ice formed (or decayed) has an isotopic composition different from seawater. The ultimate change in isotopic composition is recorded in the test of organisms which are formed by the precipitation of calcium carbonate from seawater. Finally, the remains of these fossils are deposited on the sea floor and ultimatetly buried. At each step in the transfer of information (from global climate to ice volume to ocean isotopic composition to biological activity and finally to sediment records) there is a potential for each process to alter the recorded "climate spectra". For example, as sediment particles are deposited on the sea floor, benthic organisms mix these particles with older material. In some areas of the ocean, this process may mix sediments which are several thousands of years

SUMMARY, CONCLUSIONS AND RECOMMENDATIONS 843

old with newly deposited sediments. This process, bioturbation, tend to smear the paleoclimate records of deep-sea sediments and can be described as a low-pass filter which removes the high frequency variations.

6.2. Contribution from the Symposium

At the Milankovitch Symposium, many of the aspects of estimating climate spectra were discussed and many new ideas were presented. Pestiaux and Berger reviewed several newly developed techniques of spectral analysis methods, presenting the basic assumptions, advantages and problems associated with these methods. It was shown that, although the classical Blackman-Tukey method is useful because of its statistical properties, the parametric spectral analysis methods such as the maximum entropy method or the minimum cross entropy spectral method provides an efficient tool for increasing the frequency resolution of closely spaced spectral lines and in the case that only a restricted number of data points is available. These methods were applied to the problem of non-stationary paleoclimate records. For example, Pestiaux and Berger have shown a progressive decay of the amplitude of the 100 kyr period in core V28-239 when passing from the interval 0-700 kyr BP, to the interval 900-1300 kyr BP and finally to 1300-1900 kyr BP. On the other hand, when the original data series are non-stationary and abrupt amplitudes changes are present, the Walsh spectral analysis method is recommended.

The problem of estimating a time scale for the geologic record was discussed by Herterich and Sarnthein. Previous effort to use the hypothesis of orbital forcing of climate to adjust the geological time-scales have concentrated in looking at the time-domain characteristics. In their presentation, Herterich and Sarnthein utilized the concept of spectral coherence as the mean of adjusting geologic records. They used the oxygen isotopic record from a core taken from the eastern tropical Atlantic and define an algorithm by which the time scale was adjusted so that the coherency between the geologic proxi-record and the calculated orbital tilt and precession were maximized. To define an independent time scale, other sedimentological data are used. The time scale developed by this method closely agreed with estimated time scales developed independently.

The problem of recovering the true climate spectra from geologic proxy-data was discussed in two papers (Dalfes et al. and Pestiaux and Berger). Attention was especially focused on the effects of benthic mixing processes. Dalfes, Schneider and Thompson reported on their study of effects of bioturbation on the discrete and continuous components of spectra inferred from deep-sea cores based on a simple mixing model. They concluded that the bioturbation effects can seriously distort the spectral continuum. Also, relative phases and amplitudes of the discrete components can be affected. Pestiaux and Berger presented a stable deconvolution method allowing the studying of the effect of bioturbation on the high frequency components of the geologic time series.

Finally, Morley and Shackleton discussed proxy-climate records from sediment cores taken in the same oceanic region but with different sedimentation rates. Ideally, differences between these two records should reflect difference in the sedimentation processes at each core location. They found that the differences between the geologic proxy-records varied with the parameter studied. Stable isotope records showed strong similarities between cores whereas some paleontologic data series showed marked differences in the high frequency components.

6.3. Recommendations

From the results presented at the Milankovitch Symposium and many discussions of papers, there are a number of recommendations that can be suggested in the area of estimating climate spectra from the geologic records:

1) Statistical studies of spectral analysis techniques have concentrated on the distribution of the spectral variance estimates. Efforts should be directed towards the problem of errors in both variance estimate and frequencies estimate resulting from errors in the time scale of the data series.

2) Because of the importance of time scales, efforts should be directed towards estimating an accurate geologic time scale. These efforts should be independent of the hypothesis of orbital forcing.

3) The use of orbital forcing to estimate a geologic time scale has provided some strong evidence for orbital forcing of climate. However, studies should be directed to test the pitfalls and circularity of these procedures. Numerical simulations and Monte-Carlo methods may provide useful information in this area.

4) The process of bioturbation and other post-deposital processes may have important efects on the recording of the climate spectra in deep-sea sediments. Efforts should be directed to defining the frequency response of these processes so that proxy-climate spectra can be corrected to reflect more closely the true climate record.

5) Spectra obtained through coring, sampling and measurements of deep-sea sediments are results of chain of natural and man-made processes. Each process transforms spectra. The resulting spectra are products of original climatic variable spectra with response functions of relevant intermediate processes.

Therefore, response functions of these processes should be investigated to assess their relative impacts on resulting spectra. It should be realized that a unique recovery of the original information content of climate variable time series may not always be possible. For these cases, utility of Monte-Carlo-type inversion techniques that can provide uncertainty bounds should be explored.

6.4. References

Hays J.D.,Imbrie J.,Shackleton N.J.:1976, Science, 194, pp.1121-1132
Johnson R.G.: 1982, Quaternary Research, 17(2), pp.135-147
Kominz M.A., Heath G.R., Ku T.L., Pisias N.G.: 1979, Earth Planetary Sciences Letters, 45, pp.394-410
Morley J.J., Hays J.D.: 1981, Earth Planetary Sciences Letters, 53, pp.279-295
Pisias N.G., Moore T.C. Jr.: 1981, Earth Planetary Sciences Letters, 52, pp.450-458.

7. ENERGY BALANCE MODELS (EBM) (REPORT)

Rapporteur : S.H.Schneider
Contributors : C.C. Covey, L.D.D. Harvey, T.S. Ledley, G.R. North, D. Pollard, W.D. Sellers, J.P. van Ypersele.

As the name implies, energy balance models (EBMs) determine climate by applying the law of conservation of energy. Usually this law is applied to each latitude zone on the planet, so that the rate of change of surface temperature is determined by the difference between incoming absorbed solar and outgoing infrared radiation, together with the amount of energy transported to or from adjacent latitude zones. The key simplification in this class of models is that the dynamical processes responsible for atmospheric and oceanic heat transport are completely parameterized[1] in terms of the temperature often as a simple diffusive process. This drastic approximation allows one to simulate time-evolving climates over periods of tens to hundreds of thousands of years, an impossible task for models which include explicit calculations of atmospheric or oceanic dynamics, particularly at higher resolution. A high degree of parameterization makes EBM results suspect unless they are validated by comparison with data and with the results of more comprehensive dynamical models. However, by "fitting" the model's parameters to produce a faithful simulation of today's climate, all of its degrees of freedom may be used up. Within the hierarchy of climate models, EBMs have their greatest use in determining the sensitivity of the climatic system to external forcing and internal feedback processes acting over time scales of tens of years or more. Apart from giving indications on climatic variations and variability, EBMs help to identify the main processes involved.

The Symposium contributed greatly to our understanding of the application of EBMs to Milankovitch-driven climatic change. Below, we briefly review the characteristics of different types of EBMs and summarize previous applications of EBMs to the Milankovitch problem. Then we discuss what was learned at the Symposium, and conclude this section of the report with a judgment of the most beneficial directions for future EBM research, together with an opinion as to how EBMs should interact with other areas of climate research.

7.1. Overview of pre-symposium state of the art

Shaw and Donn (1968) were the first to use a large-scale thermodynamic model to study the climate's response to Milankovitch forcing. The model they used (Adem, 1965) focused on the surface heat balance applied to the north hemisphere caloric half-years defined by Milankovitch, but did not allow the climate to vary throughout the annual cycle. Another "snapshot" approach was used by Saltzman and Vernekar (1971, 1975). Both models indicated a neglegible climatic change from Milankovitch forcing, unless large ice sheets were imposed <u>ad hoc</u>.

To simulate the dynamical nature of the climate system (e.g., the storage and release of heat by the oceans during the seasonal cycle, the growth and decay of ice sheets) one must use a time-evolving model whose components have appropriate response lags. Table 1 shows the time-dependent EBM's which have been applied to the Milankovitch problem. Among the important features of the models are the presence or absence of surface-air and land-sea differentiation and a seasonal cycle. The nature of the temperature-albedo feedback is also of critical importance in determining model behavior. Until recently, zonally symmetric models without explicit ice sheets, such as those of Schneider and Thompson (1979) and North and Coakley (1979), failed to produce temperature and ice extent changes nearly as large as those which occured during the ice ages and also produced temperature changes some 5 kyr ahead of observed changes.

Although, the response of the climate system to orbital forcing is also weak in Berger (1977), his temperature changes

Table 1 Energy Balance Models used for Milankovitch Simulations

Model	Seasons	Outgoing IR[a]	Albedo	Meridional Heat Transport	Heat Reservoirs	Longitude and Height Resolution	Ice Sheet	Time[b] (kyr BP)	Signal[b] Response
Budyko (1974)	YES	$A + BT_{SFC}$	function of T_{SFC}	Newtonian cooling	equivalent mixed layer	NONE	NO	0-230	WEAK
Sellers (1970) Berger (1977)	NO	Nonlinear function of T_{SFC}	function of T_{SFC}	linear diffusion	NONE	NONE	NO	200- -200	-- WEAK
Kallen et al. (1979)	YES	Linear in T_{SFC}	snow mass balance and ice sheet	(no latitude resolution)	deep ocean	NONE	YES	--	--
North & Coakley (1979)	YES	$A + BT_{SFC}$	function of T_{SFC}	linear diffusion	land, mixed layer	land, sea	NO	--	--
Schneider and Thompson (1979)	YES	$A + BT_{SFC}$	no albedo feedback	nonlinear diffusion	equivalent mixed layer	NONE	NO	0-100	WEAK
Sergin (1979)	NO	linearized radiative transfer	no albedo feedback	prescribed latitude profile of T	deep ocean	NONE	YES	0-600	STRONG
Suarez and Held (1979)	YES	prescribed radiative transfer	snow budget	linear diffusion	atmosphere, mixed layer	land, sea, 2 atmosphere levels	NO	0-150	MEDIUM
Pollard et al. (1980)	YES	$A + BT_{SFC}$	snow mass balance and ice sheet	linear diffusion	equivalent mixed layer	NONE	YES	0-400	WEAK

[a] T_{SFC} = Surface temperature; A, B = empirical constants
[b] Time range of forcing and amount of glacial-interglacial change obtained by models which were forced by actual insolation variations.

in high latitudes occur in the expected direction: for low obliquity, cooling extends to 55°N , which corresponds precisely to the greatest meridional extent of the ice sheet, and to 70°S; it generates also an atmospheric energy transport across all latitudes negatively correlated with temperature in high latitudes, indicating, for example, that interglacials (glacials) would be associated to a reduced (enhanced) atmospheric energy (sensible heat and water vapor) transport.

By coupling an EBM to an explicit ice sheet component, long-term fluctuations of the ice sheet were obtained which lagged the orbital forcing by about 10 kyr, due to the inertia involved in building up and reducing the massive ice sheet (Pollard et al., 1980). The magnitude of the fluctuations, about 7 degrees in latitude, corresponded to the 23 and 41 kyr period components of ice variations which are deduced from the climatic change proxies inferred from sediment data. But the dominant observed glacial-interglacial variation, at a period of 100 kyr, was largely missing from the model results. Suarez and Held (1979), with a two-layer diffusive atmospheric model, a fixed depth mixed layer coupled to a simple sea ice model and a (strong) step function temperature-albedo feedback, found a large sensitivity of the latitudinal extent of the perennial snow cover to insolation perturbation. But a phase lag discrepancy similar to that found by Schneider and Thompson (1979) was present in the model's paleoclimatic record for the last 150 000 years. One point that should be noted is that Pollard (1980) and Oerlemans and Vernekar (1981) have both shown that the feedback between the ice accumulation rate and the height of the ice sheet was more important for the sensitivity amplification than the classical albedo temperature feedback.

7.2. Symposium contributions

The primary contribution of the Symposium to EBM studies was the identification of physical processes which are good candidates for providing the link between Milankovitch insolation forcing and climate change during the ice ages. As indicated above, until quite recently EBMs had failed to reproduce a sufficiently large-amplitude, 100 kyr period variation in ice volume, even though this variation dominates the late Pleistocene geologic data. During the past year EBMs have appeared which for the first time have exhibited mechanisms for generating both the observed amplitude and frequency. Two of these models, developed by Pollard and by North, Mengel, and Short, were presented at the Symposium.

The model of North et al. provides a simple way of obtaining large glacial-interglacial fluctuations in response to the relatively small insolation changes which arise from

SUMMARY, CONCLUSIONS AND RECOMMENDATIONS

Milankovitch orbital variations. The key is seasonality and geography: this EBM incorporates realistic land-sea distribution in both latitude <u>and</u> longitude. Perennial ice is considered to cover the areas whose local temperature never exceeds freezing. For an equilibrium run with the insolation regime of 115 000 years BP, when the summer Norhern Hemisphere received less insolation than at present, ice was found to exist over a large portion of North America. In Asia, on the other hand, the relative lack of oceanic influence was such that the central land mass heated up enough in summer to melt whatever ice accumulated during the winter. The North American continent, with its much smaller east-west extent, can retain its ice during summer. In retrospect, the failure of previous EBMs to initiate reasonable ice growth can be seen, according to this theory, as a result of their lack of longitudinal resolution of the land-sea areas. Even when zonal land-sea distinction was made, as in North and Coakley (1979), they were in effect including only one continent, a "super-Asia", which efficiently melted off ice and snow in mid-summer.

The results of North <u>et al.</u> suggest that the growth of Northern Hemisphere ice sheets, when it occurs, takes place in a fairly sudden jump brought on by sufficient relative summer cooling to permit ice to survive the heating season. This idea was strengthened by results from an ice sheet model with explicit mass balance calculations which were reported at the Symposium by Ledley. In her ice sheet model, which is described in more detail in chapter 8 of this report, an initial threshold "kick" seemed to be necessary to initiate growth. The threshold for ice growth to begin seems to require both a drop in air temperature of several degrees below present and the persistence of some snow cover through the summer on land. This sensitivity of ice growth to summertime conditions on land, it was suggested, could help explain the observed tendency for lower planetary ice volume at times of high eccentricity.

With regard to the dominant 100 kyr periodicity in glacier volume inferred from the geologic record, results from coupled atmosphere-ocean-cryosphere models having internally-generated oscillatory solutions (see Chapter 9 of this report) indicate that a long time scale response - in addition to that associated with the growth and decay of massive ice sheets - can lead to a nonlinear resonance between Milankovitch forcing at 23 and 41 k$_y$r and other internal frequencies, and produce a 100 kyr cycle

The results reported by Pollard (cf. also chap. 8 of this report) from a coupled EBM-ice sheet-bedrock model show that such an internal time scale could arise from a combination of (i) the delayed response of the bedrock to the weight of the ice sheets and (ii) a hypothesized calving mechanism representing accelerated wastage at the ice-sheet tip due to the formation of proglacial lakes and/or marine incursions during deglaciations. Earlier ice-sheet models (Oerlemans, 1980, Birchfield et al., 1981) had shown that some ice-sheet retreats can be amplified somewhat by a bedrock lag of several 10^3 years, as the strongly ablating southern tip of the ice sheet retreats into the large depression left over from the formerly large ice mass. Pollard found that the addition of a calving mechanism amplified these retreats still further, producing complete deglaciations and a dominant 100 kyr power-spectral peak in the presence of the Milankovitch forcing. When the EBM-component in this model was replaced by simple prescribed snow-budget forcing, nearly the same ice-age results were still obtained. This implies that the ice-sheet component contains all the important ice-age physics, and, in this model at least, is insensitive to the precise details of the orbital forcing.

Pollard found that the best results were obtained with bedrock lag times in excess of 5000 years. This raises a paradox, for the times indicated by postglacial rebound data are closer to 3000 years. The situation was greatly clarified by the presentation of Peltier and Hyde, who pointed out, in essence, that bedrock models which have been used to date in ice sheet studies are based on an outdated model of the Earth's mantle which nevertheless yields the correct postglacial rebound rates, but for the wrong reasons. Current understanding of rock rheology implies that the entire mantle responds to the weight of large ice sheets, whereas ice sheet models have assumed that the mantle response is limited to a thin layer. Of particular significance to coupled ice-sheet-mantle models is that the mantle responds with multiple time scales because of the existence of density discontinuities at different depths. The shorter time scales are on the order of 3000 years and can account for the postglacial rebound data, but others are much longer and may be the missing ingredient needed to produce the 100 kyr cycle in the coupled models.

Pollard's model was able not only to roughly reproduce the amplitude and periodicity of the glacial-interglacial cycles, but also to follow the times series of ice volume (as implied by isotope data) with a fair degree of visual correlation. This result must be viewed with caution since it required somewhat arbitrary "fine tuning" involving, in particular, the imposition of a very fast ablation at the edge of retreating ice sheets, supposedly attributable to iceberg calving into proglacial lakes. Nonetheless, the results do indicate the potentially high sensitivity of the climatic system to such processes. In addition, all such coupled EBM/glacier models run to date have used axisymmetric glacier models to represent what, at least in the Northern Hemisphere, are three dimensional glacial state changes.

The time-evolution of the climatic system over the last deglaciation was simulated by a thermodynamic model (Adem et al.) force by the radiation values from Berger, the ice sheet boundaries by Denton and the sea surface temperature from CLIMAP. The insolation effect increases the temperature by about 1°C for 13, 10 and 8 kyr BP and 0.5°C for 7 kyr BP, with a maximum up to 3°C in continental areas like Asia and North Africa.

7.3. Opportunities for future work

The issues highlighted at the Symposium lead us to conclude that future work with EBMs should focus on at least three areas: (1) combined models in which the Earth's surface air temperature is coupled to other components of the climate system; (2) a more careful comparison of EBM results both with data and with the results of other models; and (3) improvement in heat transport parameterizations in EBMs (or the use of low-order dynamical models).

7.3.1. Coupled models

Studies using coupled EBM-ocean-cryosphere-lithosphere models should be pursued with emphasis on phase at different locations between different climatic variables and orbital forcings. The bedrock response should be modeled more realistically, with multiple time scales included which are appropriate to the scale of glacier above. Coupling of EBMs with sea ice models needs further investigation, as does coupling with deep ocean models, particularly to test the sensitivity of ice growth and decay to deep ocean transports within and between hemispheres (cf. the Symposium presentation by Harvey and Schneider). EBMs should also be used in multicomponent models to study the possible interactions between carbon dioxide variations and sea level change (e.g., Broecker's presentation), and more generally between the geochemical cycles and the climate.

7.3.2. Model Intercomparisons

7.3.2.1. With Data

A more careful evaluation of the climatic content of proxy data is also called for when EBM results are investigated. The most often used indicator of global climatic change during the Pleistocene is the oxygen isotope composition of ocean sediments, supposedly a proxy of ice volume. The presentations of Mix and Ruddiman and of Covey and Schneider point out the possibility of time lags of thousands of years between true ice volume and the marine isotope record. Furthermore, it is possible that a part of the isotope record is a response to fluctuations in the Antarctic ice cap caused by changing sea levels and the associated exposure and submergence of the Antarctic continental shelf.

7.3.2.2. With other Models

EBM results and their sensitivity to parameter changes must also be compared as much as possible with the results of other types of climate models, particularly dynamic atmospheric and oceanic general circulation models; sea ice models; and three dimensional glacier models.

7.3.3. Improving Heat Transport

More emphasis in the atmospheric component should be placed on low-resolution dynamical models; with these one might be able to treat heat transport realistically enough while keeping computer costs manageable to simulate a time-evolving climate over 10^4 years. As an example of the possible importance of poleward heat transport to the Milankovitch problem, consider the fact that eccentricity variations act mainly in the tropics. Decreased tropical heating and (perhaps) precipitation, resulting from low eccentricity, could lead to a reduction in poleward heat transport - one of fairly drastic magnitude, judging by the highly nonlinear behavior of the system which is suggested by the limited dynamical model studies to date (Thompson, 1982). Substantial ice growth could result from this amplification of a small forcing. The possibility raised here - that the ice sheets respond simply to tropical forcing - of course contradicts the more conventional view that interactions between ice sheets and other components of the climate system are responsible for the observed glacial cycles. A hierarchy of models, directed toward various interactions between the atmosphere, surface, oceans, cryosphere, lithosphere and mantle of the Earth, is needed in order to resolve this fundamental issue. One example of a high resolution EBM is Sellers' 3-D version presented to the Symposium.

In summary, energy balance models (EBMs) are a major component in the tool kit used by scientists trying to explain Pleistocene climatic variations, particularly since they are economical enough to permit simulations of a time evolving coupled air-sea-ice-land system. The primary weakness of EBMs is their lack of physical comprehensiveness, but the significance of this weakness in specific applications can often be evaluated by comparison of their short-term behavior with that of nature and of more physically comprehensive models. Their strength is practicality and ease of interpretation of cause and effect.

7.4. References

Adem, J. : 1965, Monthly Weather Review 23 (8), pp. 495-503.
Berger, A. : 1977, Palaeogeography, -climatology, -ecology 21, pp. 227-235.
Birchfield, G.E., Weertman, J., and Lunde, A.T. : 1981, Quaternary Research 15 (2), pp. 126-142.
Budyko, M.I. : 1969, Tellus 21, pp. 611-619.
Budyko, M.I. : 1974, in : "Physical and Dynamical Climatology", Proceedings of the Symposium on Physical and Dynamical Climatology, WMO n°347, Gidrometeoizdat, Leningrad.
Kallen, E., Crafoord, C., and Ghil, M. : 1979, J. Atmos. Sci. 36, pp. 2292-2303.
North, G.R., and Coakley, J.A. : 1979, J. Atmos. Sci. 36, pp. 1189-1204.
Oerlemans, J. : 1982, Climatic Change 4, pp. 353-374.
Oerlemans, J. : 1980, Nature 287, pp. 430-432.
Oerlemans, J., and Vernekar, A.D. : 1981, Contr. Atm. Phys. 54, pp. 352-361.
Pollard, D. : 1982, Nature 272, pp. 233-235.
Pollard, D., Ingersoll, A.P., and Lockwood, J.G. : 1980, Tellus 32, pp. 301-319.
Saltzman, B., and Vernekar, A.D. : 1971, J. Geophys. Res. 76, pp. 4195-4197.
Saltzman, B., and Vernekar, A.D. : 1975, Quaternary Research 5, pp. 307-320.
Schneider, S.H., and Thompson, S.L. : 1979, Quaternary Research 12, pp. 188-203.
Sellers, W.D. : 1970, J. Appl. Meteor. 9, pp. 960-961.
Sellers, W.D. : 1973, J. Appl. meteor. 12, pp. 241-254.
Sergin, V.Y. : 1979, J. Geophys. Res. 84, pp. 3191-3204.
Shaw, D.M., and Donn, W.L. : 1968, Science 162, pp. 1270-1272.
Suarez, M.J., and Held, I.M. : 1979, The sensitivity of an energy balance climate model to variations in the orbital parameters.
Thompson, S.L. : 1983, A comparison of baroclinic eddy heat transport parameterizations to a simplified general circulation model, Ph. D. Thesis, University of Washington (NCAR Co-op Thesis 71).

[1] A _parameterization_ is a parametric representation of processes occurring on scales smaller than those resolved by a model in terms of variables which are resolved. All models must parameterize small scale processes - a so-called closure approximation - details vary, of course. Another way to view parameterization is simply the statistical representation of one variable

in terms of another (e.g., outgoing infrared radiation (IR) to space in EBMs is typically parameterized in terms of surface temperature by comparing empirical data of outgoing IR with surface temperature variation with latitude or seasons).

[2] It should be mentioned here, as Pollard (1982) does himself, that Oerlemans' ice sheet model shows asymmetric glacial cycles, even with no forcing. Oerlemans' poster at the Symposium showed that the timing of the cycles was sensitive to the model parameters, while the occurrence of the long cycles was not. (See also Oerlemans, 1982).

[3] Techniques for <u>asynchronous coupling</u> of the atmospheric sub-model to other climatic system sub-models also must be developed and tested in this regard, since no dynamical coupled models can <u>continuously</u> simulate the climate for thousands of years.

8. CLIMATE MODELS WITH AN ICE SHEET

    Rapporteur : G.E. Birchfield
    Contributors : T. Ledley, D. Pollard

Some fourteen papers presented at the conference discussed models incorporating some form of an ice sheet. These papers may be classified into the following five categories.

(i) Ice and climate changes on the Milankovitch time scales concerned with data verifications (Birchfield and Weertman, Ledley, Peltier, Pollard, Oerlemans, Watts).

(ii) General circulation model studies of synoptic and seasonal climate response. (Kutzbach, Manabe, Schlesinger).

(iii) Analysis of possible fundamental interactions between major climatic components. (Ghil and Le Treut, Saltzman)

(iv) Energetics and mass flux processes at the atmosphere-ice sheet interface. (Ledley)

(v) Physics of distribution of heavy oxygen isotope ^{18}O in the environment. (Covey, Mix)

SUMMARY, CONCLUSIONS AND RECOMMENDATIONS

To focus the discussion, this summary is presented in the form of a question and answer/comments for each of the five groups.

8.1. To what extent does the physics of the ice sheets and of the underlying bedrock deformation by itself, control the climate system response to the orbital insolation anomalies, that is, to what extent does the "slow physics" in the climate system explain the observed ice sheet variations as reflected by the oxygen isotope deep-sea core record?

The considerable variance at 41, 23 and 19 kyr, observed in the proxy ice volume record has been predicted at least to a first approximation by very greatly simplified models of the ice sheets and bedrock deformation which virtually bypass the atmospheric and oceanic components (See Birchfield et al. (1981), Pollard (1982), for example). Models coupled with simple atmospheric and oceanic components have not yet significantly improved the prediction of the ice volume record (See for example, Pollard (1983)).

The large variance in the proxy ice volume record at 100 kyr has been studied intensely by ice sheet modellers. Prior to this meeting some possible mechanisms have been demonstrated including :

a) the role of isostatic adjustment and the decrease of accumulation rates with elevation as a means for rapid deglaciation (See Oerlemans (1980), Birchfield (1981), Peltier (1981), Pollard (1982), for example);

b) relatively unidentified nonlinear processes in the ice sheet physics (see Hays et al. (1976), Birchfield (1977), Imbrie and Imbrie (1980)).

Potentially important processes involved in the 100 kyr power which were presented at this meeting include :

a) ice calving either in proglacial lakes or in a marine environment as a means for rapid deglaciation. These ice-age simulations of Pollard contained large (100 kyr) cycles and agreed well with ^{18}O records over the last few 10^5 years.

b) surging, due to melting at the ice-sheet base, as a means for rapid deglaciation, can produce improved agreement with the isotope record as demonstrated in the model of Oerlemans.

c) intrinsic or self-sustained oscillations of the ice sheet coupled to an earth model, which are of sufficiently long period as to suggest possible resonant response to external forces or interaction with other components of the climate system. Oscillations of this character were identified tentatively in the model of Birchfield and Weertman, but more recent evidence suggests that these may have been spurious and that resonance response is not likely in the simple physics used.

d) an important contribution to the meeting was the demonstration by Peltier of the richness of the earth response to the varying ice load. The complex bedrock response comes from the differentiation of the interior structure of the earth into lithosphere, mantle, core, etc. Although with a realistic model of the Earth's interior, bedrock relaxation times range from hundreds to millions of years (clearly spanning the scale of the 100 kyr oscillations), Peltier's results point to the external mode, which has a relatively fast response, as being the most important. More specifically, his results suggest that for large ice sheets, the bedrock relaxation time decreases with increasing ice sheet size; this response, coupled with the incorporation of elevation damped accumulation rate at the surface of the ice sheet, when forced by snow line perturbations near 20 kyr, produces large amplitude oscillations with period near 100 kyr.

Important future work concerns :

a) further elucidation of the behaviour of ice sheet response with the improved earth response, with the other components of the system, that is, including the earth and the deep ocean.

b) the extension of Northern Hemisphere ice-sheet models to two horizontal dimensions (latitude and longitude) due to the complex longitudinal structure of the Keewatin, Hudson and Labrador topography.

c) the improved modelling and parameterization of detailed processes (e.g., ice streams, calving and basal melting) that may play a role in the rapid deglaciations.

d) modelling of marine ice shelves and grounded marine ice sheets in conjunction with the continental ice sheets.

8.2. What are the synoptic and seasonal responses of the atmosphere and ocean in the presence of large continental and possible marine ice sheets ?

The presence of the Northern Hemispheric ice sheets has large effects on the seasonal climate, mainly due to their high albedo and their high surface elevations. The primary tools

SUMMARY, CONCLUSIONS AND RECOMMENDATIONS   857

used to simulate these effects have been atmospheric general circulation models; the GCM sensitivity is found by comparing results for different ice-sheet extents and other prescribed boundary conditions corresponding to particular times in the Quaternary ice ages.

At the meeting Kutzbach presented results of one set of GCM experiments for the climate at 9 kyr BP, with particular attention to the changes in the Asian monsoon from the present. An experiment was described in which the remnant North American ice sheet at 9 kyr BP was prescribed; although this had significant effect on the northern mid-latitude climate, it had little influence on the Asian monsoon.

Manabe and Schlesinger each presented results of GCM experiments comparing conditions at 18 kyr BP with those at the present. A new mass budget calculation for the Laurentide ice sheet by Manabe, produced very large melt rates implying a long-term ice-sheet retreat far in excess of that observed. New results were presented by Schlesinger for the present and 18 kyr BP for both February and August. Sea surface temperatures and sea-ice extents were prescribed from CLIMAP data.

As far as ice-age ice-sheet modelling is concerned, it is felt that the contribution of GCMs will become very significant when simulations over a complete seasonal cycle with coupled atmospheric/oceanic GCMs becomes feasible. The net annual mass budgets over the ice sheets will then be obtainable for any era, and may be used as forcing for long-term ice-sheet integrations.

8.3. What internal oscillations can exist in the ice sheet/-earth/deep ocean system and what are possible interactions between these modes and external forcing on the Milankovitch time scale or between these modes and stochastic forcing?

Of the major climate system components, the ice sheets and the bedrock deformation offer potentially the largest response times, that is, those closest to the periods of the astronomical forcing, and have been demonstrated to be important in multiple component oscillator models. Existence of internal oscillations in simple zero-dimensional models has been demonstrated prior to this meeting (for example, see Ghil and Le Treut (1983) and Saltzman (1981)). The results presented by Ghil and Le Treut at this conference indicate, that these modes can interact with the external forcing to give response at several additional periods (combination tones) including some near 100 kyr.

8.4. What are the important surface processes operating at the atmosphere-ice sheet surface and how can they be effectively parameterized ?

As referred to above, the accurate determination of the accumulation and ablation rate at the surface of the ice sheet is a critical link in the chain between insolation perturbations and ice-sheet variations.

At the meeting Ledley made use of a surface energy, mass balance to explicitly calculate the budget; differences between this model experiment and one using a prescribed snow budget were found.

An urgent need exists for measurements of surface fluxes on the present polar ice sheets, in order to really get an effective means of parameterization of these processes.

8.5. How do processes in the free atmosphere, at the surface and inside the ice sheet control the oxygen-isotope anomaly in ice-age ice sheets?

The interpretation of paleo $\delta^{18}O$ measurements as a linear measure of global ice volume, assumes that the oxygen isotope ratio of the ice sheets is a constant. The fractionation of oxygen in snowfall tends to increase as the ice sheet grows to higher elevations, but to decrease as the ice sheet extends to lower latitudes. The studies of Covey and Mix presented at the meeting, estimated the resulting non-uniformity in anomaly in an ice sheet interior using simple models. The resultant distortion of the deep sea $\delta^{18}O$ relationship with ice volume was also estimated: lags in the deep sea core record relative to the ice volume as high as 3000 years and errors in amplitude by up to 30% could occur.

A major discrepancy has appeared between the volumes of the ice-age ice sheets as estimated from the oxygen isotope anomalies and those estimated from sea level changes on raised beach terraces. It is hoped model studies of isotopic oxygen fluxes in the climate system can lead to, for example, estimates of the anomaly in ice-age ice sheets.

8.6. References

Birchfield, G.E. : 1977, J. Geophys. Res. 82, pp. 4909-4913.
Birchfield, G.E., Weertman, J., Lunde, A.T. : 1981, Quaternary Research 15, pp. 126-142.
Imbrie, J., and Imbrie, J.Z. : 1980, Science 207, pp. 943-953.
Le Treut, H., and Ghil, M. : 1983, J. Geophys. Res. 88, pp. 5167-5190.

Oerlemans, J. : 1980, Nature 287, pp. 430-432.
Peltier, R. : 1982, Adv. in Geophys. 24, pp. 1-146.
Pollard, D. : 1982, Nature 296, pp. 334-338.
Pollard, D. : 1983, J. Geophys. Res. 88, pp. 7705-7718.
Saltzman, B., Sutera, A., and Evenson, A. : 1981, J. Atmos. Sci. 38, pp. 494-503.

9. OSCILLATOR MODELS OF CLIMATIC CHANGE

Rapporteurs : M.Ghil and B. Saltzman

9.1. Background and rationale

The observed variations of climate can be attributed to :
(i) variable external forcing, such as orbital changes;
(ii) free internal changes due to instability and non linearity in the climatic system, including both deterministic and stochastic processes, present even if external forcing were steady;
(iii) a complex combination of the above.
The situation is depicted in the schematic flow chart shown in Figure 1.

Climatic records show that a considerable amount of fluctuation of climatic variables, such as ice volume, occurs on time scales containing no known external forcing. This implies that other variables must change over such time scales. In other words, at least two climatic variables, with roughly the same time constant, must participate simultaneously in the change.

Even on time scales where some external forcing is known to be present, the observed response at that period may be so strong that a free internal variation of similar period may be required. J. Imbrie and associates have, in fact, given evidence at this symposium of a resonance at 100 kyr which is indicative of a free oscillation on this scale. More generally, if free auto-oscillatory behavior exists on any time scale, the response to a known forcing on other time scales will be influenced by the nonlinearity in the system that gives rise to the free oscillation.

The interplay of linear instability and nonlinear saturation dominates most oscillations in physical and biological, as well as climatic models. This interaction is very sensitive to

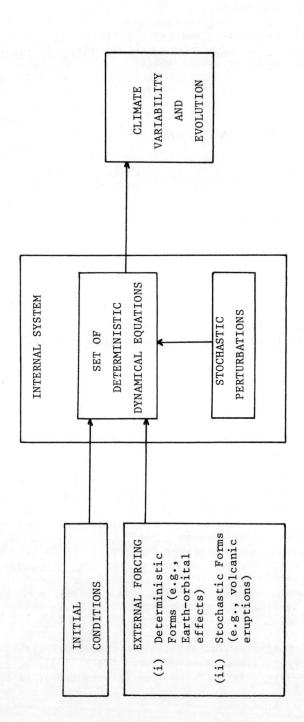

Figure 1 Schematic flow chart showing the components of a theoretical model of climatic variability (from Saltzman, 1982).

SUMMARY, CONCLUSIONS AND RECOMMENDATIONS

the presence of stochastic perturbations, which play a role therefore in climatic response.

For all these reasons it is important to explore the possibilities for free oscillatory behavior and the general effects of stochastic noise in the climatic system. Such an exploration requires the use of all the mathematical tools of dynamical system theory. The study of multiple equilibria, limit cycles, and other more complex attractor sets are part of this theory. A brief review of previous climate modeling studies along these lines is given in the next section. Contributions made at the present conference are reviewed in the third section. A summary and recommendations follow.

9.2. Previous studies

One of the earliest formulations of an "oscillator model" was made by Erikson (1968), based on a pair of equations for (i) global mean surface temperature and (ii) fraction of the earth covered by ice. He described many of the features such a system could exhibit, including the possibility for limit cycles.

An extensive development of this possibility was made by the brothers Sergin in the 1970's (Sergin, 1979 and references therein). They noted the necessity for treating two variables of roughly similar time constant (e.g., deep ocean temperature and ice mass) if one is to admit the possibility that long term oscillatory behavior of paleoclimate is of the "free" type. Further models of this kind were developed by Birchfield, Weertman and collaborators (Birchfield et al., 1981), Ghil and associates (Ghil and Le Treut, 1981), Oerlemans (1980), Peltier (1982), Pollard (1982) and Saltzman and collaborators (Saltzman et al., 1981, and references therein). All these models considered two or more fundamental climate variables: deep ocean temperature T_o and marine ice extent ℓ_m (Saltzman et al., 1981), t_o and continental ice-sheet extent ℓ_c (Källen et al., 1979), and bedrock deflection h_b (Birchfield et al., 1981; Oerlemans, 1980; Peltier, 1982; Pollard, 1982) or T_o, ℓ_c and h_b (Ghil and Le Treut, 1981).

The relevance of stochastic forcing in climate models was noted by Mitchell (1976), developed more rigorously by Hasselman (1976) and applied by Lemke (1977), and Fraedrich (1978), using a linear, one-variable model. The effects of "multiplicative coupling" of stochastic noise to such models, due to random variations of the coefficients, was studied by Nicolis and Nicolis (1979). Their work was extended by Sutera (1981), Benzi et al. (1982), and Nicolis (1982), who showed that the stochastic "exit time" for passage between two stable equilibra of a nonlinear, one-variable energy balance model is a mechanism for long term transitions in the climatic regime. Other studies of stochastic effects in climate models were made by Robock (1978), Oerlemans (1979), and North et al. (1981). Studies of the effects of stochastic noise on nonlinear, two-variable climatic systems admitting harmonic behavior were made by Saltzman et al. (1981), Saltzman (1982), and are summarized in Saltzman (1983).

The role of deterministic, periodic forcing in models exhibiting free oscillatory solutions was also considered in several studies previous to those reported at this symposium. These include the aforementioned works of Birchfield et al. (1981) and Oerlemans (1980), as well as Le Treut and Ghil (1983).

9.3. Contributions made at symposium

The presentations made at the symposium in the general area of oscillatory behavior fall into three categories :
 1) conceptual models of oscillations based on data studies,
 2) model studies of free oscillations, and
 3) model studies of forced oscillations.

W.S. Broecker presented ideas about a geochemical cycle including the chemistry of CO_2 and phosphorous, river discharge and continental shelf exposure as a result of sea-level changes. The period of such a cycle is of the order of 100 kyr ± 50 kyr. It could either cause by itself the 100 kyr peak or respond resonantly to the weak insolation forcing at this frequency.

J.Imbrie and SpecMap associates showed an analysis of phase coherence between orbital forcing and proxy data in cores with improved time scales. This analysis suggests a phase reversal in the coherence pattern near 100 kyr which is indicative of linearly or nonlinearly resonant climatic response there. An elaboration of the Imbrie and Imbrie (1980) tuned response model was proposed to explain the underlying free oscillation and its resonant behavior near 100 kyr.

G. E. Birchfield and J. Weertman outlined the posssibility of free oscillations in an $\ell_c - h_b$ ice-sheet elastic-bedrock model, as a result of the interaction between mass balance, as given by the snowline, and height of the ice sheet above sea level, as given by the bedrock deflection. The forced behavior of such an oscillator is an interesting avenue of future research. Their previous studies of forcing as well as those of Pollard, were done in a non-oscillatory regime of the model.

L.D.Harvey and S.H.Schneider presented a model in which a mixed layer was added to the deep ocean in the Källen et al. (1979) T_0- l_c. The structural stability of the original model and the modified model were studied: the self-sustained character of the free oscillations depends on parameter values and on the presence of the mixed layer. E.Battifarano and M.Ghil presented free oscillations in a similar model in which the latent heat of fusion was also included (cf. Saltzman, 1977).

C.Nicolis studied the effect of stochastic perturbations on the Ghil and Tavantzis (1983) model of T_0 and l_c. She showed that an equilibrium and a periodic solution can share stability in the presence of noise, and that residence times for either state depend on parameter values. She also showed how periodic forcing can restore phase informations, and how different models exhibiting limit cycle behavior can be reduced to certain canonical forms.

B. Saltzman and A. Sutera gave a general discussion of both free and forced oscillations. They presented also results with a forced version of their T_0- l_m model, including both deterministic and stochastic forcing. The power spectra of the forced model results showed a continuous background with superimposed peaks, like that of many proxy records of the Quaternary (cf. Mitchell, 1976). The structural instability of the system within plausible parameter ranges was also noted.

M. Ghil and H. Le Treut presented recent results with their three-variable T_0- l_c-h_b model, which has a free period of O(10 kyr). The model exhibits nonlinear resonance to the orbital periodicities due to changes in insolation (eccentricity), as well as in hydrological cycle (obliquity and precession). They predicted combination tones of the forcing periodicities, in particular near 100 kyr and 10 kyr. The latter appears to be confirmed by preliminary results on a high-resolution continental pollen core (L. Heussner and B. Molfino, this volume).

9.4. Conclusions and recommendations

The symposium was very stimulating due to the strong participation of data people, as well as model people. Progress in both observations and modeling will inevitably lead to closer interaction between the two activities. To foster such interaction is our first and major recommendation.

Dynamical evolution models, with two or more physical and chemical variables coupled by nonlinear interactions, are promising tools in explaining much of the climatic variability on Quaternary and other time scales. They are capable of self-sustained as well as forced oscillations, and of exhibiting many of the features of proxy records of glaciations such as the presence of sharp peaks on a continuous background which decays with increasing frequency (cf. Ghil and Le Treut, and Saltzman and Sutera).

These dynamic models need to be verified by comparison with other models for internal consistency, and with observational data for climatological realism. They should be compared with other simple models of the same kind (cf. Battifarano and Ghil, Birchfield and Weertman, Harvey and Schneider, and Nicolis) to study dependence of model behavior on parameter values, and on the presence of additional physical mechanisms, or of random perturbations. They should also be compared with more detailed models containing an explicit spatial variability (one dimensional and two-dimensional thermodynamic models and three-dimensional general circulation models). Intercomparisons across a hierarchy of climate models will improve both physical understanding of the phenomena and their detailed simulation.

For a comparison of dynamic, multi-variable, evolutionary models with observations there is a great need of additional data and improved analysis methods of these data. First, independent proxy indicators of various climate variables are necessary. These have to be obtained from deep-sea cores, ice cores, and continental cores. They need to include, beside $\delta^{18}O$ and planktonic foraminiferal assemblages, such novel and promising indicators as δD from ice cores, Cd for the deep-ocean circulation, diatoms and benthic assemblages from deep-sea cores, and varves and pollens from lake deposits. Some of the novel indicators would in particular give independent temperature information on the deep ocean. To verify the phase lags between climatic variables suggested by dynamic models (Bhattacharya et al., 1982; Saltzman et al., 1981), these indicators need to be dated independently of each other and of the orbital forcing.

To study the resonance phenomena and combination tones suggested by the model's oscillatory behavior, the length and resolution of the records needs to be increased. This is possible in marine deposits due to hydraulic piston cores, which might also be adapted for lake deposits. Due to observational uncertainty, greater spatial density of coring and objective, statistical analysis of the data in the time domain appears also necessary. The techniques of objective analysis developed in meteorology (Bengtsson et al., 1981) are likely to be useful for this purpose.

As for model development, these models are only at their beginning. The nature and parameter dependence of the free and forced oscillations need to be studied in detail, keeping in mind the phenomena modeled. Models incorporating all the variables and interactions, studied so far piecemeal, will have to be formulated and investigated. The effects of various types of deterministic and stochastic forcing need further attention (cf. Saltzman and Sutera, and Nicolis). The proportion of variability in the continuous part of the spectrum and in the line part needs to be established more carefully, in both models and data.

Finally, the methods and the results of this type of modeling should be extended to both longer time scales, of 10^6-10^9 yr (cf. data of Anderson, 1982), and to shorter time scales (10^{-1}-10^3 yr). It is in the latter that the practical interest resides, and there one is likely to find the ultimate payoff for all one might learn in studying glaciation cycles.

9.5. References

Anderson, R.Y. : 1982, J. of Geophysical Research 87, pp. 7285-7294.
Bengtsson, L., Ghil, M., and Källen, E. (Eds) : 1981, Dynamic Meteorology : Data Assimilation Methods, Springer Verlag, 330pp.
Benzi, R., Parisi, G., Sutera, A., and Vulpiani, A. : 1982, Tellus 34, pp. 10-16.
Bhattacharya, K., Ghil, M., and Vulis, I.L. : 1982, J. Atmos. Sci. 39(8), pp. 1747-1773.
Birchfield, G.E., Weertman, J., and Lunde, A.T. : 1981, Quaternary Research 15(2), pp. 126-142.
Eriksson, E. : 1968, Meteor. Monograph 8(30), pp. 68-92.
Fraedrich, K. : 1978, Quaterly J. of Royal Meteorological Society 104, pp. 461-474.
Ghil, M., and Tavantzis, J. : 1983, SIAM J. Appl. Math. 43, pp. 1019-1041.
Ghil, M., and Le Treut, H. : 1981, J. of Geophysical Research 86, pp. 5262-5270.
Hasselmann, K. : 1976, Tellus 28, pp. 473-485.
Imbrie, J., and Imbrie, J.Z. : 1980, Science 207, pp. 943-953.
Källen, E., Crafoord, C., and Ghil, M. : 1979, J. of Atmospheric Sciences 36, pp. 2292-2302.
Lemke, P. : 1977, Tellus 29, pp. 385-392.
Le Treut, H., and Ghil, M. : 1983, J. of Geophysical Research 88, N°C9, pp. 5167-5190.
Mitchell, J.M.Jr. : 1976, Quaternary Research 6, pp. 481-493.
Nicolis, C. : 1982, Tellus 34, pp. 1-9.
Nicolis, C., and Nicolis, G. : 1979, Nature 281, pp. 132-134.
North, G.R., Cahalan, R.F., and Coakley, J.A.Jr. : 1981, Rev. Geophys. Space Phys. 19, pp. 91-121.
Oerlemans, J. : 1979, Tellus 31, pp. 469-477.
Oerlemans, J. : 1980, Nature 287, pp. 430-432.

Peltier, W.R. : 1982, Adv. Geophys. 24, pp. 1-146.
Pollard, D. : 1982, Nature 296, pp. 334-338.
Robock, A. : 1978, J. Atmospheric Sciences 35, pp. 1111-1122.
Saltzman, B. : 1977, Tellus 29, pp. 205-212.
Saltzman, B. : 1982, Tellus 34, pp. 97-112.
Saltzman, B. : 1983, Adv. Geophysics 25, pp. 173-233.
Saltzman, B., Sutera, A., and Evenson, A. : 1981, J. of Atmospheric Sciences 38, pp. 494-503.
Sergin, V.Ya. : 1979, J. of Geophysical Research 84(C6), pp. 3191-3204.
Sutera, A. : 1981, Quaterly J. of Royal Meteorological Society 107, pp. 137-153.

10. CONCEPTUAL MODELS OF CLIMATIC RESPONSE

Rapporteur : W.F. Ruddiman

The category "conceptual models", as used in the NATO Conference, is somewhat difficult to summarize. It was used apparently to encompass models based more on descriptive aspects, but invariably with some elements of numerical underpinning. In fact, every paper in this category had some kind of quantitative treatment, including either statistical correlations or numerical parameterizations. Also, to some extent, this was the "miscellaneous" category into which papers were placed if they did not fit into others. As such, it is difficult to regard this as a distinct field of research which the conference succeeded in moving from one status to another.

Despite these constraints, there emerged several relatively coherent themes or topics, each of which drew two or three papers. These major themes are :

10.1. Insolation gradients as a climatic forcing function

Three research groups, each working separately, arrived at a similar conclusion: while orbitally-controlled changes in insolation at any one point in space are important to long-term climatic change, another factor of possible importance is changes of the _gradient_ of insolation between critical latitudes, as it was pointed out by Berger in 1976 who related it to the poleward atmospheric transport of sensible and latent heat. Young and Bradley argued that these gradient changes would directly impact the intensity of the atmospheric circulation north of 30°N, with stronger gradients causing more vigorous

SUMMARY, CONCLUSIONS AND RECOMMENDATIONS

(winter-like) circulation. McIntyre and Ruddiman suggested that the insolation-gradient changes will impact the major atmospheric circulation cells (Hadley, Ferrel) over the Atlantic Ocean and affect the net northward advection of oceanic heat from the South Atlantic toward the ice sheets; they showed one equatorial Atlantic SST curve spanning the last 250 kyr with a strong 23 kyr signal and coherence to insolation gradients reinforcing their previous findings (Ruddiman and McIntyre, 1981). Rossignol-Strick, in another session, called on insolation gradients as a control on the intensity of the African monsoon. She proposed that the intensity of this monsoonal circulation is the primary cause of freshwater run-off into the eastern Mediterranean, leading to a sequence of stagnant muds ("sapropels") over the last 250 kyr.

In short, this conference saw the emergence of an important concept, newly tested by statistical correlation to paleoclimatic data, but not yet by physical models of air or ocean circulation. This modeling phase should come next.

10.2. The role of ocean chemistry in glacial cycles

Broecker stressed the relatively symmetrical timing of the last deglaciation in the two polar regions and challenged orbital forcing as an explanation of the 100 kyr cycle. He suggested that variations in one factor common to both polar regions could explain the relatively similar response: the large CO_2 increase evident in ice cores from both Antartica and Greenland. Broecker then reviewed a published theory that calls on glacial cycles of storage and removal of organic carbon on continental shelves as a regulator of oceanic chemistry and hence of atmospheric CO_2 levels (Broecker, 1982).

For future work, more data on long-term changes in oceanic fertility and nutrient cycling are critical.

10.3. Non-linearities in the $\delta^{18}O$/ice-volume relationship

Numerous ice modelers now use $\delta^{18}O$ signals as targets for models of ice-sheet volume. This use rests on assumption that other factors in the $\delta^{18}O$ signal (local temperature, local precipitation/evaporation) can be ignored and that ice volume is dominant. Secondarily, it rests on the assumption of a constant $\delta^{18}O$ composition of glacial ice through time (-30 to -35°/₀₀). Two research groups presented papers that investigated and challenged the latter assumption.

Covey and Schneider developed a model to predict the ultimate isotopic composition of snow, given the source area for the moisture and its path or trajectory toward deposition. This model was used under present-day conditions and gave plausible results (subtropical source areas in the western North Atlantic yielding vapor that falls as -30°/₀₀ snow over the interior of Greenland). Mix and Ruddiman formulated a model of plausible ice-growth and ice-decay conditions and concluded that ice sheets are rarely at isotopic equilibrium (as opposed to mass-balance equilibrium). For a wide range of sensitivity tests, they found that this isotopic disequilibrium will result in a 1-3 kyr lag of ice volume behind the oceanic $\delta^{18}O$ signal at prominent climatic transitions.

For the future, this field needs to consider the decoupling between $\delta^{18}O$ and ice volume in spectral terms, using long time series. One significant problem is the possibility of over-representation of 41 kyr power and under-representation of 23 kyr power due to non-linearities between $\delta^{18}O$ and ice volume.

10.4. References

Berger, A.L.: 1976, Transactions of the American Geophysical Union, 57 (4), p. 254.
Broecker, W.S.: 1982, Prog. Oceanography, 11, pp. 151-197.
Ruddiman, W.F., and McIntyre, A.: 1981, Science, 212, pp. 617-627.

11. GENERAL CIRCULATION MODELS

Rapporteur : M.E. Schlesinger
Contributors : Ph. Gaspar, M. Ghil, J. Jouzel, J.E. Kutzbach, H. Le Treut, J.F. Royer, W.D. Sellers, S. Warren

11.1. Introduction

General circulation models (GCMs) simulate climate by solving a set of complex equations which are the mathematical expressions of the fundamental laws of physics - Newton's second law of motion and the conservation of mass and energy. The solution of these equations is obtained at discrete locations over the Earth's surface, at discrete levels in the vertical, and at discrete moments in time. Atmospheric GCMs (AGCMs) prerameters, atmospheric composition, land-ocean geography, and surface elevation and albedo. The sea surface temperature and sea extent must also be prescribed unless they are calculated by coupling the AGCM to models of the ocean and sea ice.

SUMMARY, CONCLUSIONS AND RECOMMENDATIONS

Because AGCMs require several hours on the fastest computers to simulate one year, they cannot yet be used to test the astronomical theory by simulating a complete 100 000 year glacial-interglacial cycle. Instead, GCMs have been used to simulate climatic "snapshots" at selected times before the present (BP) for which the required boundary conditions have either been reconstructed from geological data or assumed to be the same as at present.

The first AGCM paleoclimatic simulation was that reported by Williams (1974), Williams and Barry (1974) and Williams et al. (1975) for the last glacial maximum. The NCAR[1] six-layer AGCM with 5° latitude-longitude resolution was used to simulate the January and July climates for both modern and ice age surface boundary conditions obtained from a wide variety of sources. The simulations showed decreased ground temperature and precipitation, but increased low-level cloudiness, over North America and Europe in January and July during the ice age; a displacement of the zones of cyclonic activity to the south of the major ice sheets, but no corresponding shift in the jet stream; and a replacement of the Asian monsoon low by high pressure.

The ice age and modern July climates were also simulated with a two-layer, 4°x5° latitude-longitude resolution AGCM by Gates (1976a,b) and with the 11-layer, 265 km uniform resolution GFDL[2] model by Manabe and Hahn (1977), both with the ice age boundary conditions as reconstructed by the CLIMAP[3] Project Members (1976) for 18 kyr BP[4]. These models' simulations gave global-mean surface air temperatures that were colder during glacial maximum than at present (4.9°C and 5.4°C, respectively), with larger decreases over ice-free land (5.8°C and 6.2°C) than over open ocean (2.7°C and 2.4°C), and reductions in the global-mean precipitation rate (14% and 10%, respectively). The study by Gates (1976a,b) showed a southward displacement and intensification of the jet stream in the vicinity of the major ice sheets, along with an equatorward shift in the zones of maximum eddy activity, and a weaker Asian monsoon. A weaker ice age monsoon was also reported by manabe and Hahn (1977) who showed that it resulted from the larger cooling over land than over ocean. This was caused by the prescribed increase in continental albedo rather than by the prescribed changes in sea surface temperature.

Two simulations of the climate 9 kyr BP have been performed to determine the effects of changes in the earth's orbital parameters. A brief account of the earlier simulation of June climate with the 5-layer UKMO[5] AGCM coupled to a 2-meter deep ocean was given by Gilchrist and Mitchell (see Mason, 1976). The later simulation of June-July-August (JJA) and December-

January-February (DJF) climates with a 5-layer, 11°x11° latitude-longitude resolution AGCM with prescribed modern surface boundary conditions was reported by Kutzbach (1981) and Kutzbach and Otto-Bliesner (1982). They found that the 7% increase in solar radiation during JJA 9 kyr BP induced an intensified Asian monsoon as a result of high continental surface temperature and decreased pressure relative to those over the surrounding ocean. These changes were accompanied by increased precipitation and precipitation minus evaporation in qualitative agreement with the geological evidence of enlarged paleolakes. In DJF the 7% decrease in 9 kyr BP solar radiation was found to result in weaker changes of the opposite sign to those of JJA.

11.2. Symposium Papers

Manabe and Broccoli investigated the influence of the 18 kyr BP ice sheet on climate using the GFDL[2] AGCM coupled to a 68.5 m deep ocean/sea ice model. They obtained simulated climates from each of two versions of the model: one with the present land ice distribution, and the other with the 18 kyr BP land ice distribution as reconstructed by CLIMAP. In the Northern Hemisphere, the sea surface temperature (SST) difference between the two experiments resembled the SST difference between 18 kyr BP and present as reconstructed by CLIMAP. Only a very small difference in model SST was found in the Southern Hemisphere. This suggested that other changes in the Earth's heat balance, besides those caused by changes in ice sheet distribution and Milankovitch insolation variations, are needed to fully explain the global ice age climate.

Kutzbach presented results from a new simulation of the 9 kyr BP climate obtained by the 9-layer, 5°x7° latitude-longitude resolution NCAR[1] Community Climate Model (CCM), and compared these results with those obtained previously from a lower-resolution GCM. The patterns of the simulated climate changes were nearly the same for both models, but the magnitude of the response was increased for the CCM experiment. This demonstrated the need to perform intermodel comparisons of GCM paleoclimatic simulations.

Royer presented simulations obtained by a 10-layer, 9°x11° latitude-longitude resolution AGCM with orbital parameters for 125 kyr and 115 kyr BP (Berger, 1978), but with contemporary surface boundary conditions. Although the differences in the annual-mean global-mean climatic variables simulated for 125 kyr and 115 kyr BP were small, there were larger changes in the seasonal cycles with a marked tendency for cooler summers and weaker monsoon circulations over the northern hemisphere continents at 115 kyr BP. At this time the simulated annual-mean temperature and soil moisture over eastern Canada were colder and wetter, respectively, than at 125 kyr BP, a possible indication that insolation changes alone may have led to favorable conditions for the inception of the Laurentide ice sheet.

Sellers presented climate simulations for the present, 9 kyr and 231 kyr BP obtained by a 5-layer, 10°x10° latitude-longitude resolution model with simplified dynamics, an interactive ocean and the seasonal cycle of insolation as computed from Berger (1978). It was found that the precipitation rate simulated for 9 kyr BP was larger than that simulated for the present, and that the pattern of increased precipitation was similar to that obtained by Kutzbach. Also, the simulated seasonal cycle was larger 9 kyr BP, and smaller 231 kyr BP, than today.

11.3. Recommendations for Future Work

The recommendations of the Panel are presented below under four principal activities.

11.3.1. Analysis of Existing Paleoclimate Simulations

For the 231 kyr, 125 kyr, 115 kyr, 18 kyr and 9 kyr BP "snapshot" climate simulations which have been made it is recommended that :

- The level of statistical significance of the paleoclimatic change be determined;
- The statistically-significant climatic quantities be compared with the climatic data to validate the models;
- The simulations from several GCMs and simpler models be intercompared to establish the level of model dependence of the results;
- The statistically-significant, model-independent results be analyzed to understand the mechanisms of climate change, for example, whether the snow mass budget agrees with the history of the advances and retreats of the ice sheets.
- The cycles of trace elements for which modern and paleodata exist, as the isotopic species of water (HDO and $H_2^{18}O$) or the aerosol content, be incorporated in the models. This can, in turn, be also helpful to validate the models when applied to paleoreconstruction.

11.3.2. Development of Coupled Atmosphere/Ocean GCMs

To simulate paleoclimates the SST and sea ice extent should be predicted rather than prescribed and the reconstructed fields used to validate the models. Therefore, it is recommended that :

- AGCMs be coupled with models of the ocean and sea ice;
- Such atmosphere/ocean (A/O) GCMs be tested by comparison of their modern and 18 kyr BP climate simulations with the observed modern and reconstructed data;
- The model errors be analyzed and corrected.

11.3.3. Simulation of Selected Times During the Last Glacial-Interglacial Cycle

To test the astronomical theory with GCMs it is recommended that :

- "Snapshot" simulations of the annual cycle be conducted with the validated A/O GCMs for the time of glacial onset and at 3 kyr intervals over the last glacial-interglacial transition from 18 kyr BP to the present;
- The paleoclimatic data necessary for boundary conditions (e.g., ice sheet extent) and validation be assembled for the above times;
- The model dependence of the simulations be determined.

11.3.4. Development of Coupled Atmosphere/Ocean/Ice Sheet GCMs

To simulate a complete glacial-interglacial cycle it is recommended that :

- Coupled models of the atmosphere, ocean and ice sheets be developed;
- Accelerated integration techniques such as asynchronous coupling of the component models be developed to make a 100 kyr simulation practical.

Finally, it is suggested to create a catalogue of paleoclimatic data of interest for the modelers :

- input data, as boundary conditions (sea-ice and glacier extents, SST, albedo ...) or data potentially useful in the model itself (CO_2 or other trace gases contents, dust load of the atmosphere ...);
- output data; ground level temperature, precipitation rate, atmospheric circulation and windspeed changes, relative humidity over the sea-surface ...

An effort has to be made to quantify the data, to obtain values representative of large geographical sectors and to concentrate the paleoclimatic reconstruction over the above mentionned periods.

11.4. References

Berger, A. : 1978, J. Atmos. Sci. 35(12), pp. 2362-2367.
CLIMAP Project Members : 1976, Science 191, pp. 1131-1137.
Gates, W.L. : 1976a, Science 191, pp. 1138-1144.
Gates, W.L. : 1976b, J. Atmos. Sci. 33, pp. 1844-1873.
Kutzbach, J.E. : 1981, Science 214, pp. 59-61.
Kutzbach, J.E., and Otto-Bliesner, B.L. : 1982, J. Atmos. Sci. 39, pp. 1177-1188.
Manabe, S., and Hahn, D.G. : 1977, J. Geophys. Res. 82, pp. 3889-3911.
Mason, B.J. : 1976, Quart. J. Roy. Meteor. Soc. 102, pp. 473-498.
Williams, J. : 1974, Simulation of the atmospheric circulation using the NCAR global circulation model with present day and glacial period boundary conditions. University of Colorado, Boulder, 328pp.
Williams, J., and Barry, R.G. : 1975, in : "Climate of the Arctic", G. Weller and S.A. Bowling (Eds), University of Alaska Press, pp. 143-149.
Williams, J., Barry, R.G., and Washington, W.M. : 1974, J. Appl. Meteor. 13, pp. 305-317.

[1] National Center for Atmospheric Research, Boulder, Colorado.

[2] Geophysical Fluid Dynamics Laboratory/NOAA, Princeton University, Princeton, New Jersey.

[3] Climate : Long-Range Investigation and Prediction.

[4] 18 kyr BP indicates 18 000 years before the present.

[5] United Kingdom Meteorological Office, Bracknell, Berkshire, United Kingdom.

AUTHORS INDEX

Ablowitz, M.J., 498
Adam, D.P., 382, 387, 543, 558, 673
Adelseck, C.G.Jr., 200
Adem, J., 95, 528
Adhemar, J., 12, 83
Aharon, P., 380, 382-385
Ahmed, N., 422
Akaide, H., 432
Alexander, R.C., 738
Alvarez, L., 253
Alvarez, W., 253
Anderson, E.J., 164
Anderson, R.Y., 80, 152, 154, 156, 159-160, 164, 167
Anderson, T.F., 200, 214
Andrews, J.T., 234, 546, 602, 673, 711, 715
Anolik, M.V., 13, 17, 40, 44
Arhelger, M.E., 237
Armstrong, R.L., 143
Arnold, V.I., 67, 639
Arrhenius, G., 200
Arthur, M.A., 166, 178, 187, 194-197, 199, 203, 206-208, 210-214
Asaro, F., 253
Ascencio, J.M., 514, 689, 694, 798
Auffret, G.A., 206, 208

Backer, P.A., 212, 543
Bacon, M.P., 332, 421
Barnett, T.P., 405, 450
Barrow, E.J., 84, 582
Barry, R.G., 227, 285, 673, 711
Bausch, W.M., 198, 200
Be, A.W.H., 242-243, 349, 542, 674, 681
Beauchamp, K.G., 422, 430, 562
Belov, N.A., 225, 241-242
Bengtsson, L., 66, 81
Benninger, L.W., 270, 450
Benson, L.V., 81, 136
Bentley, C.R., 560

Benzi, R., 242, 543, 638
Berger, A.L., 3-4, 12-14, 17, 19, 21-25, 35, 37, 55, 58, 72, 84-85, 86, 88, 95, 103-111, 113-117, 124, 144, 170, 178, 183-185, 187, 192, 224-226, 234, 236, 274, 276, 277, 317, 340-341, 380, 382-383, 385, 386, 410, 413, 418-419, 423, 436, 448, 460, 478, 495, 522, 527-528, 543, 545, 551-552, 566, 568, 574, 578, 582, 607, 616, 637-638, 674, 677, 690, 696, 708, 716, 734-735, 738, 766, 780, 802, 804, 817
Berger, W.H., 196, 200, 203, 209-210, 256, 332, 482, 482, 484
Berggren, W.A., 183, 373
Bernabo, J.C., 241, 393
Bernard, E., 37, 88, 460
Berner, W., 689, 798, 803
Bhanu Kumar, O.S.R.U., 350, 362
Bhattacharya, K., 562, 655, 657
Birchfield, G.E., 183, 225, 514, 524, 542-543, 546, 560, 567-568, 570, 572, 574-576, 584, 600, 602, 607
Birkhoff, G.D., 14, 66
Bishop, J.K.B., 260, 332
Bishop, T.N., 256, 421, 431
Blachut, S.P., 229, 231
Blackman, R.B., 421, 560
Blackmon, M.L., 803, 803-804
Blasco, S.M., 231-232
Blifford, I.H.Jr., 332, 421
Bloom, A.L., 380-381, 431
Bloutsos, A.A., 582, 727
Blytt, A., 164, 167
Bolli, H., 237, 260
Bornhold, B.D., 232, 232
Bosellini, A., 167, 180
Bottero, J.S., 349-350

Boulanger, M., 727, 735
Boulton, G.S., 678, 683
Bourke, W., 464, 736, 803-804
Bowman, K., 736, 796
Box, G.E.P., 423-425, 464, 495
Bradley, R.S., 84, 84
Bradley, W.H., 148, 164, 167, 178, 186, 344
Brakenridge, G.R., 338, 421
Branstator, G.W., 803-804
Braslau, N., 285, 717
Brenner, G., 203, 259
Bretagnon, P., 13-14, 17-18, 43-46, 48-49, 52, 58, 572
Broecker, W.S., 224, 237, 241, 285, 297, 344, 380-382, 482, 553, 568, 584, 674, 688, 697, 798
Brouwer, D., 6-7, 10-13, 17, 58-59, 64, 67, 570
Brown, J., 85, 225, 256, 380, 382, 410
Bruckner, E., 83-84, 803
Bruckner, W.D., 180, 195
Bruder, K.F., 227, 229, 256
Brumberg, V.A., 43, 717
Bruns, T., 423, 450, 460-461, 464
Bryan, K., 655, 666
Bryson, R.A., 332, 349, 542, 708
Budd, W.F., 425, 542, 562
Budnikova, N.A., 13, 450
Budyko, M.I., 519, 674, 678, 766, 780
Bukner, D.H., 150, 450
Bunker, A.F., 666, 678
Burckle, L.H., 201, 203, 331
Burg, J.P., 332, 405, 407, 422-423, 431-432
Burgers, J.M., 542, 560
Burrell, D.C., 232, 237
Buys, M., 12, 37, 605, 678

Cadzow, J.A., 423

Cahalan, R.F., 513, 515, 566, 617
Calder, N., 225, 689
Callegari, A.J., 659
Canada Department of Transport, 225, 227
Canadian Government Travel Bureau Data, 717
Canetti, H.J., 735
Carlson, T.N., 201
Cathles, L.M., 560
Cayan, D.R., 673
Cepek, P., 195, 198-199, 203-204, 208
Cess, R.D., 717
Chamley, H., 285
Chandrasekhar, S., 624
Chappell, J., 86, 380-382, 386, 542, 564, 673, 718
Chapront, J., 43
Chebotarev, G.A., 5
Chervin, R.M., 745, 805
Chiu, L.S., 778
Chlonova, A.F., 259
Ciesielski, P.F., 248
Clapp, P.F., 528
Clark, D.L., 242, 681
Clarke, G.K., 431
Clausen, H.B., 692, 695, 701
Clemence, G.M., 6-7, 10-13, 58-59, 64, 67
Climap, 308, 331, 349, 392, 527, 529, 791
Coackley, J.A.Jr., 116, 513, 515, 566, 617, 716-717
Cobban, W.A., 168, 188, 208
Cochran, J.K., 246
Coiffier, J., 736
Cole, K., 780
Coleman, M., 214
Colette, B.J., 560
Compston, W., 382
Cooley, J.W., 507
Corezzi, S., 168
Cornet, W.B., 143, 178, 187
Covey, C., 701
Cox, A.V., 188, 208, 567
Crafoord, C., 569, 638

AUTHORS INDEX

Crary, A.P., 224, 227
Crawford, G., 653-655, 657, 662, 666
Croll, J., 12, 84, 182, 192, 270
Crowley, T.J., 566, 674-675
Crowley, W.P., 497
Crutcher, H.L., 792-793
Curry, W.B., 352
Curtis, C., 214

Dalrymple, G.B., 274, 285, 419, 448, 567
Danielson, E.F., 332
Dansgaard, W., 237, 542, 692, 695, 701
Dauphin, J.P., 311
Dave, J.V., 717
Davis, M.B., 393
De Boer, P.L., 37, 164, 172, 181, 199-200, 208, 260
De Graciansky, P.C., 206, 208
De Moor, G., 736
Dean, W.E., 152, 178, 195, 198-201, 203-204, 206, 208, 265
Defant, A., 185
Delany, A.C., 201
Delaunay, A., 67
Delibrias, G., 270, 285
Delmas, R.J., 514, 689, 694, 798
Denton, G.H., 224, 527-528, 546, 560, 683, 805
Deprit, A., 67
Deque, M., 95, 712, 735, 818
Dey, B., 362
Dodge, R.E., 270
Doell, R.R., 248
Donn, W.L., 673-674
Doppert, J.W.C., 375
Douglas, R.G., 203, 210, 212, 256, 260
Dreimanus, A., 691
Duing, W., 350, 352
Dunn, D.A., 25, 86, 183, 548, 556
Dupeuble, P., 206, 208
Duplessy, J.C., 234, 270, 285, 349, 450, 462-463, 494, 498, 502
Duprat, J., 270
Durazzi, J.T., 234, 756
Duriez, L., 18, 37, 43, 45, 48, 109

Eder, W., 200
Edmond, J.M., 332
Eicher, V., 692
Einarsson, T., 248
Einsele, G., 164, 178, 256, 260
Ekdale, A.A., 210, 212
Elderfield, H., 212
Eliasen, E., 519-520, 522, 524
Ellis, D.B., 310
Ellwood, B.B., 248
Emiliani, C., 140, 270-271, 273, 275, 385, 547, 673, 687
Epstein, S., 381
Erdélyi, A., 71
Erlenkeuser, M.H., 448, 450-451
Ernie, Y., 736
Evenson, A., 569, 615, 618-619, 624, 628, 638
Ewing, M., 673, 688

Fairbanks, R.G., 270
Fairbridge, R.W., 729
Fatschel, J., 198
Fillon, R.H., 224-225, 234, 238
Fischer, A.G., 37, 164-166, 168, 172-173, 178, 187, 194, 199, 204, 206-208, 252, 261
Flohn, H., 708

Folger, D.W., 201
Fong, P., 633
Forsythe, G.E., 72, 75-76
Fougères, P.F., 422
Fox, P.T., 723-724
Frakes, L.A., 151
Francou, G., 43
Frerichs, W.E., 260
Fryrear, D.W., 332
Fuji, N., 405-406
Fukuo, Y., 406

Gardner, J.V., 178, 195, 198-201, 203-204, 206, 208, 265, 334-335, 674
Garrison, R.E., 200, 208, 212
Gaspar, Ph., 95, 497
Gates, W.L., 311, 331, 338, 343-344, 350, 770-771, 773-774, 776-777
Gavin, J., 778
Gedzelman, S.D., 701
Geiss, J., 673
Geitzenauer, K.R., 285, 331, 349
Ghil, M., 12, 37, 80-81, 494, 543, 553, 569, 605, 616, 638-639, 641, 651, 653-655, 657, 662, 666
Gieskes, J.M., 212
Gilbert, G.K., 136, 164, 167, 173
Giles, B.D., 727
Gillette, D.A., 332
Glaccum, R.A., 332
Glass, B.P., 496
Glen, J.W., 584
Glover, L.K., 496
Goldberg, E.D., 201, 332
Goldstein, H., 58-59, 67
Goodman, J., 281
Goodwin, P.W., 164
Gordon, A.L., 448
Gordon, C.T., 789
Goreau, T.J., 482, 496
Gow, A.J., 381

Griffin, J.J., 201, 332
Grosswald, M.G., 224, 227, 232, 683
Grove, A.T., 349, 802
Guetter, P.J., 95
Guinasso, N.L., 208, 482, 484, 497
Guiot, J., 85, 183, 448, 460
Gundestrup, N., 692, 695

Haagenson, P.L., 301, 701
Hahn, D.C., 311, 338, 607
Hahn, D.G., 350, 362, 607, 789
Hallam, A., 195, 200, 204
Hallam, B., 200, 259
Hambelen, C., 198, 259
Hamilton, A.C., 136, 527
Hammer, C.U., 659, 692, 695
Hammond, S.R., 332, 335
Hannan, E.J., 85, 425
Hansen, A.R., 615, 625, 633, 789
Hansen, J.E., 625, 767
Harland, W.B., 167, 188, 208, 248
Harvey, L.D.D., 275, 656, 659, 661, 666-667
Harzer, P., 12-13, 209
Hasselmann, K., 301, 450, 482, 487, 494, 643
Hattersley-Smith, G., 86, 227
Hattin, D.E., 192, 198, 204, 248
Hayder, E., 372, 514
Hayder, M., 600, 767
Hays, J.D., 3, 25, 84, 86, 140, 148, 151-152, 182, 192, 200-201, 238, 252, 256, 275-276, 281, 297, 307, 312, 314, 332, 334, 335, 338, 339, 342, 344, 353, 355, 358, 362, 391, 395, 397-400, 402, 418, 447-448, 454, 456, 458, 460, 467-469, 481, 487, 542

489-491, 548, 554, 566,
568, 578, 605, 674, 683,
692, 699, 711, 789, 801
Heath, G.R., 86, 209, 273-
275, 280, 297, 301, 307,
311-312, 331, 338-339, 418,
447-448, 484, 487, 495,
548, 561
Heezen, B.C., 201, 259, 688
Held, I.M., 513-514, 519,
524, 542, 561, 796, 801
Heller, P.C., 168, 278
Herman, Y., 242-244, 248, 560
Herterich, K., 13, 85
Heusser, C.J., 335, 381
Hide, W., 527, 605
Hinkle, R.T., 244, 278
Hite, R.J., 150, 406
Hochuli, P.A., 199, 204, 259
Hoffert, M.I., 656, 659
Hofman, D., 198, 259
Holloway, J.L., 692, 789
Honjo, S., 253, 332
Hooghiemstra, H., 371-372,
519, 607
Hope, G.S., 381, 387
Hopkins, D.M., 244, 244, 248
Hori, G., 67, 527
Horie, S., 387, 406
Houghton, D.D., 789, 804
Hsieh, C.T., 659, 692
Hsu, K.J., 227, 253
Hughes, T.J., 224, 527-528,
546, 560, 683, 796, 805
Hunkins, K., 242-243, 546,
681
Hutson, W.H., 349, 450, 496

Iida, K., 248, 406
Imbrie, J., 3, 25, 80, 84-86,
111, 124, 140, 148, 182,
192, 200, 225, 238, 252,
256, 260, 270, 273-275,
278, 280-282, 288, 297,
307, 314, 339, 344, 353,
355, 358, 362, 380, 383,

386, 391, 395, 397-400,
402, 410, 418, 447-448,
454, 458, 472, 481, 487
489-491, 514, 542-543, 548,
554, 566-568, 578, 605,
633, 674, 683, 689, 692,
699, 711, 734, 789, 801,
789, 801
Imbrie, J.Z., 85, 225, 274,
278, 297, 362, 410, 454,
514, 543, 548, 568, 633,
683, 689
Imbrie, K.P., 80, 109, 192,
200, 260, 270, 383, 386,
514, 801
Ingersoll, A.P., 331, 542,
561
Ingle, J.C.Jr., 192, 248, 331
Irvin, H., 214, 248
Irving, G., 542, 692

Jackson, M.L., 332
Jaenicke, B., 332
James, R.W., 723-724
Janecek, T.R., 328, 333, 335,
337-338
Jansa, L.F., 178, 195, 199,
201, 203-204, 208
Janssen, C.R., 393
Jantzen, R., 674
Jenkins, G.M., 183, 278, 280,
318, 421, 423-425, 448, 495
Jenkyns, H.C., 195, 200
Jenne, R.L., 792-793
Johnsen, S.J., 692, 695, 701
Johnson, L.R., 332, 337
Johnson, R.F., 210
Johnson, R.G., 273-275, 297,
301, 543, 548, 551, 671,
673, 681
Johnson, R.W., 436
Jones, G.A., 482
Jones, P.D., 723
Jouzel, J., 743

Källen, E., 81, 569, 638,

653-655, 657, 662, 666
Kanari, S., 406-408
Karlin, K., 159
Kauffman, E.G., 168, 204, 206-207
Kellogg, T.B., 676
Kelly, P.M., 724
Kemp, E.M., 262
Kennedy, W.J., 188, 200, 208, 212
Kennett, J.P., 256
Ketten, D.R., 332
Killingley, J.S., 210, 214
King, T.A., 310
Kinney, P., 237
Kipp, N.J., 566-567, 674
Kipphut, G., 285, 482
Kira, T., 406
Kirkland, D.W., 149, 152
Kolmogorov, A.N., 67
Kominz, M.A., 86, 273-275, 280, 297, 301, 307, 312-313, 339, 447-448, 487, 495, 548
Komro, F.G., 666
Koopmann, B., 463
Koppen, W., 83-84
Krasheninkov, V., 260
Krassinsky, G.A., 13, 17, 44
Kremp, O.W., 257
Krissek, L.A., 332
Ku, T.L., 86, 241, 244, 273-275, 280, 297, 301, 307, 312, 339, 418, 447-448, 487, 495, 548
Kukla, G.J., 3, 85, 95, 183, 225, 297, 380, 382, 410, 448, 460, 484, 486, 778-779
Kukla, H.J., 779
Kullenberg, B., 270
Kutzbach, J.E., 95, 136, 185, 187, 338, 350, 361-362, 708, 734, 744, 765-766, 780-781, 783, 802-806, 818

Lacis, A.A., 767

Lacoss, R.T., 423
Lagrange, J.L., 10
Laird, A.R., 701
Lamb, H.H., 362, 673, 708, 710, 727
Land, L.S., 214
Langway, C.C.Jr., 701
Lapina, N.N., 241-242
Laplace, P.S., 10
Larsen, S.E., 410, 495
Larson, J.A., 224
Laseski, R.A., 393
Laursen, L., 519, 522, 524
Lawrence, J.R., 701
Lebel, B., 694
Ledbetter, M.T., 248, 502
Ledley, T.S., 585, 703
Leetma, A., 352
Le Goff, G., 736
Legrand, M., 514, 689, 694, 798
Leinen, M., 310-312, 331, 468
Lemke, P., 481-482
Lepas, J., 736
Lepas, J., 736
Le Treut, H., 80, 494, 543, 553, 616, 638, 651, 655, 657
Le Verrier, U.J.J., 10, 12-13
Lewis, C.F.M., 232
Lian, M.S., 717
Lie, S., 67
Linkova, T.I., 242
Livingston, D.A., 136
Llewellyn, P.G., 188, 208
Lloyd, C.R., 185
Lockwood, J.G., 542, 561
Lorenz, E.N., 80
Lorius, C., 743
Lotti, R., 85, 225, 380, 382, 410
Lowenstam, H.A., 381
Lozano, J.A., 692
Lunde, A.T., 524, 542-543, 546, 560, 567-568, 570, 572, 574-576, 584, 607
Luz, B., 200, 203, 331
Lyons, J.B., 237

AUTHORS INDEX

Mackey, S.D., 681, 681
Mahaffy, M.A.W., 234, 310, 673, 711, 715
Makrogiannis, T.J., 721, 727
Malcolm, M.A., 72, 75-76, 308
Malone, R.C., 793, 803-804
Manabe, S., 311, 338, 350, 362, 395, 566, 666, 734, 789, 791
Mankinen, E.A., 274, 285, 419, 448, 715
Manspeizer, W., 134, 136
Marple, S.R., 419, 422, 432
Martinson, D.G., 25, 84, 86, 192, 314, 353, 355, 358, 362, 395, 395, 397-398, 402, 418, 548, 605
Mathieu, G., 224, 242-243, 681, 715
Matter, A., 203, 208, 210, 212
Matthews, R.K., 224, 226, 234, 270, 285, 344, 380-382, 447-448, 460
Maurasse, F., 270, 285
Maykut, G.A., 422, 583
McCann, S.B., 224, 231
McCave, I.N., 178, 199, 206, 208, 715
McClure, B.T., 448, 671, 673, 681
McCune, A.R., 76, 134, 143
McIntosh, W.C., 143, 143
McIntyre, A., 25, 84, 86, 159, 192, 200, 224, 234-236, 307, 314, 338-339, 343, 353, 355, 358, 362, 391, 395, 397-398, 402, 418, 467-468, 514, 543, 548, 558, 569, 605, 674-676, 678-679, 692, 712, 735, 756, 805
McKenzie, J., 199, 199
McKinney, C.R., 380-381
McLaughlin, D.B., 137, 143

Meier, R., 148, 285
Mengel, J.G., 285, 514, 516, 586, 818
Mercer, J.H., 212, 224, 231, 691
Merlivat, L., 738, 743
Meserve, J.M., 673, 792-793
Mesolella, D.J., 285, 380
Mesolella, K.J., 344, 380-382, 715
Michel, H.V., 253, 338
Middleton, P., 701
Milankovitch, M.M., 12-13, 83-85, 88, 95, 111, 114, 120, 192, 224, 254, 270, 343, 391, 395, 566, 616, 674, 677, 801
Minhinick, J., 432, 721, 723
Miskovitch, V.V., 13, 285
Mittelberger, J.P., 681, 736
Mittra, R., 423, 727
Mix, A.C., 25, 84, 86, 192, 314, 342, 353, 355, 358, 362, 395, 397-398, 402, 418, 548, 605, 678, 681, 704
Mobley, R.L., 736, 738
Moder, C.B., 72, 75-76, 308
Molfino, B., 159, 192, 285, 349
Molina-Cruz, A., 310, 331, 334, 691
Monecchi, S., 168, 242
Montadert, L., 206, 208, 210
Mook, W.G., 342, 394-397, 402
Moore, P.D., 226, 393
Moore, T.C.Jr., 25, 86, 183, 308, 310-311, 314, 319, 331, 338, 342, 358, 487, 548, 551, 556, 605
Moore, W.S., 231, 234
Morgan, K.A., 671, 681
Morgan, V.I., 394, 703
Moritz, R.E., 88, 227, 638, 655, 657, 666
Morley, J.J., 25, 84, 86, 192, 273-276, 280, 282,

288, 297, 307, 312, 314,
334-335, 339, 342, 353,
355, 358, 362, 391, 395,
397-398, 402, 418, 447-448,
456, 458, 460, 469, 548,
605
Morse, I.E., 278, 285
Mosby, H., 231, 231
Moser, J., 64, 67, 183
Muller, C., 206, 208, 210
Murphy, J.J., 83-84, 419
Musson Genon, L., 736, 736

Namias, J., 673
Napoleone, G., 168
Natland, J.H., 206
Neftel, A., 689
Neibler-Hunt, V., 234, 756
Neugebauer, J., 212
Newell, R.E., 387, 671, 778
Newton, I., 4
Nicholson, S.E., 708
Nicolis, C., 80, 605, 638, 643-644, 647
Nicolis, G., 647
Ninkovitch, D., 226, 311, 338
Nishimura, S., 406
North, G.R., 116, 513-516, 566, 586, 617, 818

Obradovich, J.D., 168, 174, 208
O'Connor, M.J., 232
Odin, G.S., 167-168, 173, 188
Oerlemans, J., 183, 514, 542-543, 546, 553, 560, 566, 570, 575, 600, 602, 607-608, 703
Oeschger, H., 689, 798
O'Keefe, I.A.A., 253, 263
Olausson, E., 270
Olsen, P.E., 129-131, 134-136, 138-139, 141, 143, 164, 167, 178, 187

O'Neil, J.R., 242, 244, 248
Oort, A.H., 675
Opdyke, N.D., 225-226, 242-243, 248, 270, 273, 285, 288, 334-335, 338, 353, 372, 419-420, 447-448, 450-451, 460, 494, 502, 507, 548, 551, 566-567, 607, 676, 681, 710
Osmond, J.K., 244, 246
Ostlund, H.G., 299
Otto-Bliesner, B.L., 185, 187, 350, 361-362, 734, 744, 802-806, 818

Padgham, R.C., 333
Paggi, L., 168, 178, 181, 203
Paltridge, G., 724
Pannella, G., 187
Pantic, N., 251, 257, 259, 262
Parizi, G., 543, 638
Parkin, D.W., 201, 332-333
Parzen, E., 425
Pastouret, L., 199-200, 203, 285
Paterson, W.S.B., 572, 602
Peltier, W.R., 560, 570, 572-573, 575-576, 605
Peng, T.H., 285, 482
Perch-Nielsen, K., 203, 210, 212
Perruzza, A., 201
Pestiaux, P., 17, 25, 58, 85-86, 95, 124, 183, 418-419, 423, 436, 448, 460, 495, 497-498, 712, 818
Petersen, E.L., 410, 495
Peterson, G.M., 338
Peterson, J.A., 387
Pflaufmann, U., 463
Pickton, C.A.G., 188, 208
Pierrard, M.C., 736
Pietreck, H., 332
Pisias, N.G., 25, 84, 86, 183, 192, 273-275, 280,

AUTHORS INDEX

282, 288, 297, 301, 307, 310-314, 323, 338-339, 342, 353, 355, 358, 362, 395, 397-398, 418, 447-448, 467-468, 487, 495, 548, 551, 556, 605
Pitcher, E.J., 803-804
Pius, L.J., 13, 17, 44
Poincaré, H., 66
Pollard, D., 514, 542-543, 546, 550, 558, 560-562, 564, 584-585, 602, 607, 654, 665
Poor, H.W., 495
Porter, S.C., 691
Pratt, L.M., 199, 206-207
Prell, W.L., 25, 84, 86, 192, 200, 273, 280, 282, 285, 288, 314, 349, 352-353, 355, 358, 362, 395, 397-398, 418, 468, 548, 605, 674, 802
Premolli, S.I., 168, 178, 181, 203
Price, P., 310
Prospero, J.M., 201, 332-333
Pujol, C., 270
Puri, K., 803-804

Quay, P.D., 299

Ramanathan, V., 803-804
Rao, K.R., 422
Rea, D.K., 328, 333, 335, 337-338
Reek, N., 692, 695
Reid, J.L., 308
Reimnitz, E., 227, 229
Reinmann, B.E.F., 201
Remington, C.L., 178, 187
Renaud, D., 694
Ricken, W., 198
Rivals, S., 736
Robertson, J.H., 308, 311, 314, 331, 334-335
Rocafort, J.P., 736
Rochas, G., 736
Rochas, M., 736
Rogers, J.C., 227
Rooth, C.G.H., 495, 798
Rossignol-Strick, M., 349, 367
Rousseau, D., 736
Rovelli, A., 432
Royer, J.F., 95, 712, 735-736, 818
Ruddiman, W.F., 86, 159, 200-201, 224, 234-236, 307, 338-339, 343, 358, 387, 391-392, 467-468, 482, 496, 514, 543, 558, 569, 599, 674-676, 678-679, 681, 692, 704, 712, 735, 756, 805
Ruelle, D., 80

Sachs, H., 260, 331
Sacks, V.N., 188, 241-242
Saltzman, B., 3, 513, 553, 569, 615-616, 618-619, 624-625, 628, 633, 638, 655, 657, 666
Sancetta, C., 237, 308, 331, 496
Sander, B., 164, 708
Sarnthein, M., 85, 252, 338, 448, 450-451, 463
Sattinger, D., 569, 649
Savin, S.M., 214, 214-215, 237, 256, 260
Scheibnerova, V., 254, 260
Scheidegger, R.F., 332, 332
Schink, D.R., 208, 210, 482, 484
Schink, N.D., 496-497
Schlanger, S.O., 210, 212, 282
Schneider, S.H., 513, 542, 586, 638, 656, 659, 661, 666-667, 745, 805
Schnitker, D., 450, 487, 735

Scholle, P.A., 203, 206, 208, 210, 212, 214, 252
Schramm, C., 448, 468
Schroder, C., 448, 448, 450-451
Schutz, C., 463, 770-771, 773-774, 776-777
Schutz, L., 331-332, 337
Schwarzacher, W., 37, 164-166, 178, 208, 252, 666
Sciarrillo, J.R., 282, 487
Segur, H., 497-498
Seibold, E., 178, 195, 198-199, 201
Sellers, W.D., 95, 513, 519, 566, 674, 717
Semtner, A.J., 436, 583
Sergin, V.Ya., 543, 655, 735
Shackleton, N.J., 25, 84, 86, 140, 148, 182, 192, 200, 203, 225-226, 234, 238, 248, 252, 256, 273-275, 280-282, 285, 288, 297, 307, 312, 314, 334-335, 338-339, 344, 353, 355, 358, 362, 372, 380, 386, 391, 395, 397-400, 418-420, 447-448, 450, 451, 454, 458, 460, 469, 481-482, 487, 489-491, 494-495, 502, 507, 542, 548, 551, 554, 566-568, 578, 605, 607, 674, 676, 683, 692, 699, 701, 710-711, 789, 801, 701, 710-711, 789, 801
Shannon, C.E., 431, 583
Shapiro, M.A., 701
Sharaf, S.G., 13, 24
Sharp, R.P., 252, 381
Shaw, D.M., 674, 681
Shaw, H.F., 86, 496
Shen, W.C., 241, 708
Shore, J., 423, 436, 735
Short, D.A., 514, 516, 586, 656, 818
Shukla, J., 280, 362
Siddiqui, M.M., 25, 86
Siegenthaler, U., 692, 770

Silker, W.B., 308, 332
Simon, J.L., 43, 708
Simpson, G.C., 674, 681
Sinicin, V.M., 238, 254, 262
Smith, A.G., 130, 167, 188, 208, 352
Smith, G.I., 149, 487
Smith, I.N., 542, 562, 735
Smith, R.L., 349-350
Smyrlie, D.E., 431, 436
Snider, H.I., 152, 487
Solomon, A.M., 393, 569
Solomon, J.A., 393, 649
Spears, D.A., 310, 770
Spelman, M.H., 666, 770
Start, G.G., 25, 86, 735
Stauffer, B., 689, 770
Stefanovic, D., 234, 251
Stern, W.F., 789, 789
Stockwell, J.N., 12, 24, 735
Stokes, W.L., 673, 681, 735
Stouffer, R.J., 771, 789, 798
Stow, D.A.V., 206, 701
Street, F.A., 256, 338, 349, 656
Streeter, H.F., 273, 280, 282, 288, 352-353, 661
Streeter, S.S., 215, 234, 238
Stuiver, M., 299, 487
Stuiver, S., 381, 381
Suarez, M.J., 513, 519, 542, 561, 616, 796, 801
Sutera, A., 543, 569, 615, 618-619, 624-625, 625, 628, 633, 638
Swain, A.M., 338, 349
Swander, J., 689, 770
Syers, J.K., 332, 332

Taljaad, J.J., 792-793
Tamburi, A.J., 237
Tauber, H., 542, 701
Tavantzis, J., 80, 638-639, 641
Taylor, K.E., 111
Tetzlaff, G., 463

Thiede, J., 285, 331
Thierstein, H.R., 285
Thom, B.G., 718
Thomas, R.H., 560
Thompson, K.S., 134, 143, 178, 187
Thompson, P.R., 282, 331
Thompson, S.L., 513, 542, 582
Thurber, D.L.J., 285, 344, 380, 382
Till, R., 310
Tisserand, F., 9
Tolderlund, D.S., 387
Tong, H., 422
Tse, F.S., 278
Tukey, J.W., 421
Turon, J.L., 270
Twenhofel, W.H., 141

Ulrich, T.J., 421, 431
Urey, H.C., 381
U.S. navy hydrophic office, 528

Van Andel, T.H., 180, 331
Van Blaricum, M.L., 423
Van Campo, E., 349
Van Donck, J., 224, 285, 297, 344, 372, 380, 553, 566-568, 688
Van Erve, A.W., 257
Van Graas, G., 180
Van Hinte, J.E., 168, 180, 194, 208
Van Houten, F.B., 130-132, 136, 138-139, 148, 178
Van Loon, H., 792-793
Van Woerkom, A.J.J., 13, 17, 674
Van Ypersele, J.P., 95
Van der Hammen, T., 338
Veeh, H.H., 380, 542, 564
Vergnaud-Grazzini, C., 199-200, 203

Vernekar, A., 12, 513, 780
Volat, J.L., 199-200, 203
Von Grafenstein, R., 448, 450-452
Von Zeipel, H., 66
Vonder Haar, T.H., 675
Vulis, I.L., 655, 657
Vulpiani, A., 543, 638

Walcott, R.I., 546
Walsh, J.L., 428
Walter, H., 393
Walters, R., 188, 208
Wang, C.C., 25
Warren, S.G., 586
Watts, D.G., 183, 278, 280, 318, 421, 448
Watts, R., 514
Webb III, T., 338, 392
Weertman, J., 183, 225, 278, 514, 524, 542-543, 546, 560, 567-570, 572, 574-576, 584, 599-600, 607
Wegener, A., 83-84
Weissert, H., 199
Welch, P.D., 421
Wenkam, C., 331
West, D.G., 393
Wetherald, R.T., 566, 734
Weyl, P., 671
Wigley, T.M.L., 569, 724
Wijmstra, T.A., 338
Wilcock, B., 167
Williams, D.F., 224-225, 234, 238, 349
Williams, L.D., 711, 756
Williams, M.A.J., 338
Wilson, C., 199, 201
Windom, H.L., 311, 331-333, 337
Winterer, E.L., 180
Witman, R.R., 681
Woillard, G.M., 392-397
Wolter, K., 463
Wonders, A.A.H., 37, 164, 172, 178, 180-181, 208

Woodroffe, A., 673
Woodruff, F., 260
Woodruff, S., 724
Wu, P., 576
Wyrtki, K., 350

Xu Qinqi, 86

Yamamoto, A., 406
Yang, I.C., 381
Yokoyama, T., 406
Young, M.A., 84

Zagwijn, W.H., 375
Zumbrunn, R., 689

SUBJECTS INDEX

Accumulation rates, 467-468, 480
Accuracy, 3, 103, 299
Advection
  warm water, 678
Albedo
  feed-back, 367, 678
  initiator, 671
Anhydrite, 150-151
Anticyclonic conditions, 747
Appennines, 195
  rhythmically bedded Cretaceous pelagic limestones, 195
Arabean Sea, 349
Arctic Ocean, 223-224, 241
  Canadian, 715
  ice sheet, 223
  oceanic-climatic regime, 248
Aridity, 367-368
Astronomical
  elements, 83, 109
  forcing, 85
  influences, 185
  parameters, 87
  periodicities, 109
  signals, 84, 103
  solutions, 4, 14-16, 25, 35, 37, 83, 103
  theory, 3-4, 83-85
Atmospheric circulation, 331, 708

Basal-water formation, 609
Bedding thickness, 164, 208
  frequencies, 208
Bedrock sinking, 607
Beds, 164
Bifurcation, 599
BIM boundaries, 297
Biosphere
  effect of the -, 256
Bioturbation, 209-210, 323
Black shale, 150
Bottom water
  cooling, 238

  residence time, 208
Boundary
  Triassic-Jurassic -, 143
Brunhes-Matuyama magnetic reversal, 269-270, 274
Bundles, 164, 170

C. davisiana, 467-468, 470-473, 476, 479
CaCo3, 197
  dissolution, 197
Calcite, 151
Calcium
  carbonate, 155-156, 160
  sulfate, 155-156, 160
California, 149
Caloric equator, 95, 177, 184-185
Calving, 541-544, 546-549, 552-555, 557-558, 560-562
Carbonate, 177, 181
  compensation depth, 180
  cycles, 192, 203
  diagenetic "unmixing", 200
  dilution, 200
  dissolution, 200-201, 212
  production, 200
Castile Formation, 151, 154, 157
Celestial mechanics, 4, 7, 10, 24, 41
Cementation
  $\delta^{18}O$, 203
Chemical compaction, 212
Chronology, 384
Circulation, 185
  intensity, 186
Clay-mineral assemblage, 206
Climate, 185
  - cooling, 715
  monsoonal -, 349
  simulation of -, 527
Climatic
  - belts, 259, 262
  - change, 615
  - evolution, 607

- forcing, 155
- regimes, 86, 242
Climatic cycles
  stable isotope, 206
Climatostratigraphic, 187
Coherence spectra, 317
Coherency, 296, 298, 467, 478-479
  spectrum, 297, 301
Cold summers, 743
Colombia (bore hole of funza), 371
Colorado, 150
Continental drift, 607
Continentality, 516
Cooling, 610
Coral reef climatic record, 381
Coring gap, 282
Correlations, 155
Correlations coefficients, 472, 479
Cretaceous, 177-178, 208
  cycles, 208
  limestone-marlstone sequences, 200
  lower to middle -, 193
  terrestrial vegetation, 259
  tune, 208
Cross-correlation, 472-473
Cross-spectral analysis, 296, 317, 447-448, 460
Crystalline quartz, 311
Cycle, 382
  100-ky -, 297
  chemical -, 131, 136, 138
  $\delta^{18}O$, 203
  detrital -, 131-132, 136, 138
  Eocene, 203
  hemicycle, 382
  limit -, 616, 625, 628, 635
  lockatong -, 138
  pre-& post Jaramillo glacial -, 604
  productivity, 203
  sedimentary -, 130
  site 366, 203

Sr/Ca, 203

$\delta^{18}O$, 223, 234, 301, 380
Deep-sea sediment cores, 307
Delaware basin, 151-152
Dephasing, 367
Depositional energy
  cyclic variations, 200
Diagenetic
  Sr/Ca ratio, 214
  $\delta^{18}O$, 214
  overprints, 210
  potential, 212
  unmixing, 200
Digital filters, 278, 294, 299
Dilution
  site 502 Caribbean, 201
  Walvis ridge, 201
Dilution (terrigenous -), 201, 206
  climatic cycles, 206
Dilution and dissolution, 201
Distribution of insolation, 113
Dolomite, 150
DSPD site 366, 195

Earth's orbital elements, 3-4, 11, 17, 24, 35, 37, 85
  accuracy, 3-4, 14, 35
  stability, 3
Eccentricity, 3-4, 14, 20, 24, 26-27, 32, 35-36, 86, 163, 170, 386, 566
  cycles, 148-149, 155, 183
Ecliptic, 6
Effect of orbital variations, 780
Energy balance models, 513, 519
Energy balance net budget, 582
Eocene cycles

radiolarian, 203
SiO_2, 203
SiO_2/Al_2O_3, 203
upwelling, 204
Eocene, 178
  carbonate cycles, 203
  DSPD site 366, 203
  green river formation, 148
Eolian, 331
Equilibrium line net budget, 585
Equinoxes, 88
Eustatic fluctuations, 729
Evaporites, 148-149

Foraminifer
  planktic -, 352
Foraminiferal
  biostratigraphy, 242
Forcing
  eccentricity -, 616, 622, 628, 633
  long-period -, 615
  random -, 609-610
  stochastic -, 616
Formations
  late Triassic Lockatong -, 130, 143
  Passaic -, 130, 136, 143
Fourier coefficients, 114
Fourier representations, 113
Fourier series, 116, 124
Frequencies, 3, 18, 24, 35, 86, 103, 109
  accuracy, 103, 109
  geological data, 86
  insolation, 86
  orbital elements, 3, 24, 35
  stability, 24, 103, 109
Frequency
  distribution, 476
  domains, 25, 86, 109
  spectra, 124
Freshening, 156, 159
Function
  elliptic, 67

Hamiltonian -, 58
Funza (Colombia), 371, 375
  - record, 372, 377
  - stage numbers, 372

General Circulation Models, 734, 789, 802
General planetary theory, 18, 37
Geodynamic phenomena, 251-252, 254
Ghil's model, 639, 643
Gilbert unit, 187
Glacial
  bottom-water, 237
  cycle, 372, 607
  nucleation, 715
  onset, 715
Glacial-interglacial
  transitions, 734
Glaciation, 671
Glacierization (instant), 756
Glacio-eustatic, 151
Grande Pile, 391, 404
Great basin, 136
Great lakes of Africa, 136
Greenhorn formation, 198, 204
  dilution, 204
  pelagic carbonate, 204
  redox and carbonate
    dilution cycle, 204
Gubbio
  average bedding
    periodicities, 210
Gyre
  North Pacific, 308
  Northern subtropical, 675

Halite laminae, 150
Halite, 151
Holocene, 199

Ice sheet, 225, 541-544, 546, 548-549, 553, 556, 558-561, 564, 607, 678, 683,

805
  accumulation, 225
  behavior, 683
  climate, 789
  dynamics, 607-609
  Laurentide - cap, 756
  Laurentide -, 715
  model sensitivity, 585
  model, 607
  northern Eurasian -, 223
  three-phase - accumulation, 225
  three-phase - accumulation, 225
Ice,-703, 701-702
  - albedo feedback, 513
  - caps, 379
  - shelves, 223, 232
  - volume, 565
  continental -, 380
Index
  orbital insolation monsoon, 367
Influence of 18 kyr BP, 789
Initiation of glacial cycles, 599
Insolation gradient, 707
  atmospheric circulation, 708
  feedback mechanisms, 712
  ice volume record, 711
  moisture transport, 712
  paleoclimatic comparisons, 708
Insolation, 3, 83, 108, 156, 170, 367, 379, 566, 729
  accuracy, 83, 85-86, 103
  calendar date, 87
  curve, 478
  daily, 87-88
  latitudinal, 85, 89
  mid-month, 87, 89-95, 103, 104-108
  monthly mean, 87, 95, 98-102
  monthly values, 84
  seasonal, 85, 89, 566
  spectrum, 83, 86, 94-103, 108

stability, 85
variation, 735
Interglacials, 715, 802, 817
  120 kyr BP, 817
  9 kyr BP, 802
  glacial-interglacial transitions, 734
  last -, 733
Interhemispheric correlations, 691
Isostatic adjustment, 565, 570
Isotopes
  Uranium series -, 242
Isotopic substage, 735
Italy, 177

Jaramillo event, 282
Jurassic, 149
  - terrestrial vegetation, 257
  Jurassic-Cretaceous, 193

Lacustrine, 186
Lag, 386
Lagrange
  equation, 4, 7, 11-12
  method, 11
Lake Biwa, 405, 413
Laminated black claystone, 193
Land-sea geography, 515
Late Cretaceous-Paleogene, 203
  Gubbio, 203
Late Eocene, 195
Late Pleistocene, 317, 467-468
Latitudinal gradients, 84
Leads, 386
Limestone, 154, 181
  limestone-marlstone alternations, 198
Limit
  Poleward limit of

SUBJECTS INDEX

continents, 599
Linear system, 278
Long period
  inequality, 10, 17
  theories, 44
Long-term variations, 3, 11-12, 20, 36-37
Longitude of the perihelion, 4, 20, 36
Low latitudes, 379
Lysocline, 203

Maiolica limestone, 163
Marls, 181, 214
  Hauterivian-Barremian, 214
  North Atlantic, 214
Meltwater, 541, 558-559, 561
Method
  Euler -, 75
  of averages, 69
  predictor-corrector -, 75
Milankovitch insolation variations, 608, 610
Milankovitch theory, 24, 84, 95, 163, 367, 384, 565, 687, 729
Miocene (upper)
  cycles, 199
  DSDP site 532, 199
Miocene, 195, 203, 265
  carbonate cycles, 203
  chalk-marl cycles, 195
Model
  Andrews-Mahaffy -, 715
  Coupled -, 616
  Ghil's -, 639, 643
  North-Coakley -, 514
  Saltzman -, 638
  Seasonal climate -, 519
  Seasonal energy balance -, 513
  Thermodynamic -, 527
Model, 541-546, 548, 552-558, 560-562
  Thermodynamic -, 527
Modelling, 542

orbital variations -, 715
Moisture
  flux, 671
  initiator, 671
Monsoon, 185, 367, 765, 802
  Indian -, 349
  Precipitation, 811
  Southwest -, 364
Multivariate regressive models, 85

New Mexico, 149, 151
Newark basin, 130, 135
Newark supergroup, 129
North Atlantic
  surface waters, 674
North-Coakley model, 514
Northwest Pacific Ocean, 308
Nutrients, 186

Obliquity, 3-4, 14, 20, 21, 24, 28-29, 33, 35-36, 86-87, 89, 95, 159, 163, 183, 223-224, 274, 478, 565-566
Ocean
  model, 653, 655-657, 661, 665-666
  sediments, 699, 701, 704
Opal content, 323
Opaline silica, 310
Orbit
  osculating elliptical, 5, 7
  planetary, 6-7
Orbital
  eccentricity, 386
  element changes, 513
  elements, 60
  forcing, 307, 391, 688
  parameters, 185, 515, 737
  periods, 164
  tuning, 286, 313
Organic-carbon rich layers, 199

Oscillations
  damped -, 606
  internal -, 608-609
  self-sustained -, 605
Oscillator, 637-638
  anharmonic, 67
Oxygen isotope
  curves, 478
  data, 382
  measurements, 467
  ratio, 468
  records, 473, 476
  signal, 479

Paleoceanographic, 331
Paleocene, 195
Paleoclimatological, 331
Paleosol, 186
Palynological record, 371
Paradox, 150
Parameterization of the large-scale heat transport, 519
Pelagic carbonate sediments
  burial, 210
  cementation, 210
  compaction, 210
  dissolution, 210
Pelagic carbonate, 192
  late Cretaceous-Paleogene, 203
  rhythmically bedded, 203
Pelagic sequence, 177, 193, 199
  Cretaceous Tethyan, 199
Period, 3, 86, 565
  2700-year -, 154
Periodicities, 265, 675
Permian, 151
Perturbation
  - of the coordinates, 58
Perturbation function (also disturbing function), 4-5, 8-9, 11-13, 17
Perturbations
  general, 7-8

mixed, 8
periodic, 8-9
secular, 8, 13
special, 7
Phase
  lags, 278
  relationships, 472
  spectra, 300
  spectrum, 299
Phytogeographic realms, 259
Planetary system, 3-4, 37
Planetary theories, 43
Plateau
  Tibetan -, 350
Pleistocene glacial cycles, 610
Pleistocene, 149, 331
Pliocene, 265
Pollen, 391, 401
Precession, 3, 14, 20, 22-24, 30-31, 34-35, 86, 89, 95, 148, 159, 163, 170, 177, 183, 223-224, 274, 323, 325, 479, 565-566
Precessional cycle, 150, 155, 386
Primary depositional origin lithologic criteria, 198
Problem
  - of three bodies, 56
  Kepler -, 56
  N-body -, 66
  reduced three-body -, 56

Quartz content, 325
Quasiperiodicities, 83
Quaternary ice-ages, 83
Quaternary, 200, 265
  variations in productivity, 200

Radiation
  solar -, 349
Radiolarian

assemblage, 325
fauna, 310, 323
Radiometric
  age, 282
  methods, 380
Rank curve, 140
Red spectrum, 313
Redox cycle, 193, 199
Reflux, 156
Regime(s)
  Arctic oceanic climatic -, 248
Resonance, 299
Response
  non-linear, 323, 325
Response to variations in the orbital parameters, 519
Rhythms
  Barremian, 164
  bedding, 163-164, 170, 177, 195
  bundle, 164
  Cenomanian, 164
  climatic, 163

Sapropels, 367,
Scaglia Bianca limestone, 164,
Scaglia formation, 195,
Scales,
  magnetic polarity time- -, 242,
  radiometric -, 143,
Scisti a Fucoidi marls, 164,
Sea ice model sensitivity, 591, 595,
Sea-ice, 223, 768,
  cover, 671,
Sea-surface temperature estimates, 467,
Sea-surface temperature, 468, 470-473, 476, 479, 721,
Searles lake, 149, 476, 479, 721,
Seasonal, 476, 479, 721,
  climate model, 519, 476, 479, 721,
energy balance models, 513, 476, 479, 721,
harmonics, 515, 476, 479, 721,
insolation contrast, 382, 476, 479, 721,
insolation, 113-114, 476, 479, 721,
Seasonal behaviour, 84, 476, 479, 721,
Seasons, 476, 479, 721,
  astronomical, 87, 95-96, 479, 721,
  caloric, 97, 566, 479, 721,
  half-year astronomical, 87, 89, 479, 721,
  half-year caloric, 87, 95, 479, 721,
  meteorological, 87, 95, 479, 721,
Sedimentation, 185, 479, 721,
  - rates, 210, 479, 721,
Sensitivity study, 733, 479, 721,
Series, 479, 721,
  Laplace -, 78, 479, 721,
Sharp transitions, 688, 479, 721,
Single-exponential system, 299, 479, 721,
Small divisors, 17, 479, 721,
Snow, 479, 721,
  - cover, 349, 479, 721,
  - line, 608, 479, 721,
Solar constant, 4, 87, 768, 721,
Solstices, 88, 768, 721,
Solutions, 768, 721,
  aperiodic, 79, 768, 721,
South Atlantic, 467-468, 768, 721,
Spain, 768, 721,
  pelagic carbonate, 195, 768, 721,
Specmap age model, 472, 768,

721,
Spectra, 298, 467, 479, 567,
  cross- -, 467, 479, 567,
Spectral analysis, 3-4, 25,
  32-34, 86,
Spectral response, 674, 32-
  34, 86,
Spectrum, 32-34, 86,
  analysis, 317, 32-34, 86,
  coherence, 317, 312, 86,
  red, 313, 86,
  variance, 313, 86,
Stable oxygen isotope, 214,
  86,
Stacking, 286, 86,
Stages, 86,
  boundaries, 273, 86,
Statistical significance of
  the model response, 745,
  86,
Stochastic, 86,
  component, 608, 86,
  forcing, 616, 86,
Stratification, 185, 86,
Stylolitic contacts, 195, 86,

Subtropical highs, 768, 86,
Surface energy budget, 816,
  86,
Surface productivity, 208,
  86,
Surface-water salinity, 206,
  86,
Symmetry, 156, 159, 86,

Temperature, 379
Termination, 687-688, 691
  major glaciations, 687
  southern hemisphere, 694
  synchroneity in the two
    hemispheres, 691
Terrestrial plants, 258-260,
  263
Tertiary, 178
Texas, 151
Time

constant, 278, 298
domain, 86, 109
scale, 209, 299, 312
Todilto, 149
Tradewinds, 331
Transient response, 656, 658-
  659, 666
Transition (between climatic
  regimes), 607-608
Triassic, 178
  - lockatong formation, 148
Tropical ocean, 380
Tuning
  orbital, 313, 323
Two-body problem, 5, 6

Umbrian Appennines of Italy,
  164
Upwelling
  coastal -, 349
  monsoonal -, 361
Utah, 150
Uvigerina peregrina, 238

Valleys
  Rift -, 129
Variables
  action-angle -, 67
  canonical -, 58
Variance, 298, 476
  spectra, 313
Variations
  - of constants, 58
  eccentricity -, 605, 615
Varve, 148, 150-151
  interpretation, 156
Varves (couplet)
  non-glacial -, 141
Volcanism, 256-257, 259-260
Volume change, 201
  dilution, 201
  dissolution, 201

SUBJECTS INDEX

Walvis Ridge, 199
Ward Hunt Ice Shelf, 223, 227
Water depth
  cyclic variations, 200
Westerlies, 331
Western interior of North
    America, 198